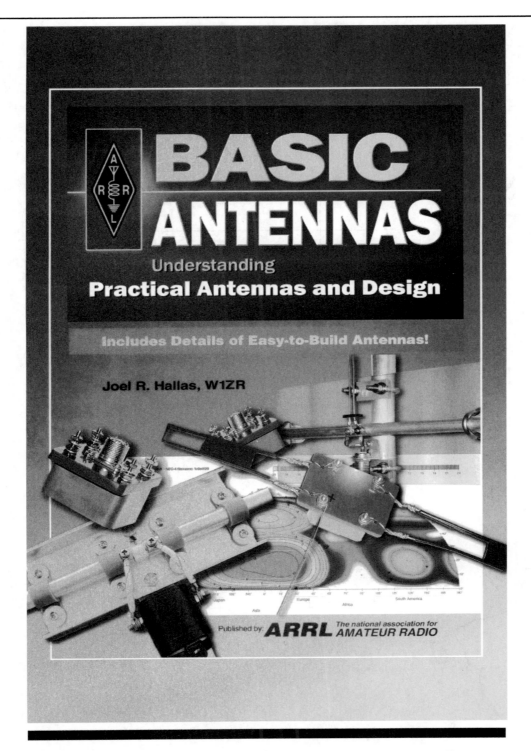

BASIC ANTENNAS

Understanding
Practical Antennas and Design

Includes Details of Easy-to-Build Antennas!

Joel R. Hallas, W1ZR

Published by: **ARRL** The national association for AMATEUR RADIO

Author
Joel R. Hallas, W1ZR

Production
Michelle Bloom, WB1ENT
Carol Michaud, KB1QAW
David Pingree, N1NAS

Cover
Sue Fagan

Foreword

Antennas are one of the most critical elements of a radio system, yet perhaps one of the least understood. Most people involved in the radio arts enter from a background in the circuitry required to define radio equipment and many have difficulty making the transition to the electromagnetic side of radio.

This book is intended to assist those with a basic knowledge of radio technology in making that important leap from the circuit domain to the antenna domain. The technology is developed using the minimum of mathematical concepts to allow introduction of basic principles in an easy to read manner. This book assumes that you have been through and understand the concepts presented in *Understanding Basic Electronics* or a similar course of study in basic electronics principles. It also is assumed that you understand electronics as applied to basic radio systems as was presented in *Basic Radio — Understanding the Key Building Blocks*.

Upon completion of this book, readers should have enough understanding of the basic principles of antenna systems to be able to easily make decisions about selection of antennas for their applications or proper use of more advanced materials on the topic, such as are found in *The ARRL Antenna Book*.[1] There you will find the detailed numerical and mathematical data required to successfully construct and deploy many of the antenna types described here.

As with all ARRL books, be sure to check to see if there are any last minute changes that didn't get into the book before it went to the printer. Updates and errata, if any, can be found at **www.arrl.org/product-notes/**.

David Sumner, K1ZZ
Executive Vice President
Newington, Connecticut
October 2008

---

[1] R. D. Straw, Editor, *The ARRL Antenna Book*, 20th Edition. Available from your ARRL dealer or the ARRL Bookstore, ARRL order no. 9043. Telephone 860-594-0355, or toll-free in the US 888-277-5289; **www.arrl.org/shop/**; **pubsales@arrl.org**.

Table of Contents

The national association for Amateur Radio

The seed for Amateur Radio was planted in the 1890s, when Guglielmo Marconi began his experiments in wireless telegraphy. Soon he was joined by dozens, then hundreds, of others who were enthusiastic about sending and receiving messages through the air—some with a commercial interest, but others solely out of a love for this new communications medium. The United States government began licensing Amateur Radio operators in 1912.

By 1914, there were thousands of Amateur Radio operators—hams—in the United States. Hiram Percy Maxim, a leading Hartford, Connecticut inventor and industrialist, saw the need for an organization to band together this fledgling group of radio experimenters. In May 1914 he founded the American Radio Relay League (ARRL) to meet that need.

Today ARRL, with approximately 150,000 members, is the largest organization of radio amateurs in the United States. The ARRL is a not-for-profit organization that:
• promotes interest in Amateur Radio communications and experimentation
• represents US radio amateurs in legislative matters, and
• maintains fraternalism and a high standard of conduct among Amateur Radio operators.

At ARRL headquarters in the Hartford suburb of Newington, the staff helps serve the needs of members. ARRL is also International Secretariat for the International Amateur Radio Union, which is made up of similar societies in 150 countries around the world.

ARRL publishes the monthly journal *QST*, as well as newsletters and many publications covering all aspects of Amateur Radio. Its headquarters station, W1AW, transmits bulletins of interest to radio amateurs and Morse code practice sessions. The ARRL also coordinates an extensive field organization, which includes volunteers who provide technical information and other support services for radio amateurs as well as communications for public-service activities. In addition, ARRL represents US amateurs with the Federal Communications Commission and other government agencies in the US and abroad.

Membership in ARRL means much more than receiving *QST* each month. In addition to the services already described, ARRL offers membership services on a personal level, such as the ARRL Volunteer Examiner Coordinator Program and a QSL bureau.

Full ARRL membership (available only to licensed radio amateurs) gives you a voice in how the affairs of the organization are governed. ARRL policy is set by a Board of Directors (one from each of 15 Divisions) elected by the membership. The day-to-day operation of ARRL HQ is managed by a Chief Executive Officer.

No matter what aspect of Amateur Radio attracts you, ARRL membership is relevant and important. There would be no Amateur Radio as we know it today were it not for the ARRL. We would be happy to welcome you as a member! (An Amateur Radio license is not required for Associate Membership.) For more information about ARRL and answers to any questions you may have about Amateur Radio, write or call:

ARRL—The national association for Amateur Radio
225 Main Street
Newington CT 06111-1494
Voice: 860-594-0200
Fax: 860-594-0259
E-mail: **hq@arrl.org**
Internet: **www.arrl.org/**

Prospective new amateurs call (toll-free):
800-32-NEW HAM (800-326-3942)
You can also contact us via e-mail at **newham@arrl.org**
or check out *ARRLWeb* at **http://www.arrl.org/**

Chapter 1

Introduction to Antennas

This impressive collection of antennas is part of the ARRL station, W1AW

Contents

So What is an Antenna?

For something so simple to check out, and often so simple to make, an antenna is remarkably difficult for many people to understand. That's unfortunate, because for many radio systems the antenna is one of the most important elements, one that can make the difference between a successful and an unsuccessful system.

How Can You Picture Antennas?

Perhaps an analogy from *The ARRL Antenna Book* will help. You are familiar with sound systems. Whether represented by your home stereo or by an airport public address, sound systems have one thing in common. The system's last stop on its way to your ears is a *transducer*, a device that transforms energy from one form to another, in this case — a loudspeaker. The loudspeaker *transforms* an electrical signal that the amplifier delivers into energy in an acoustic wave that can propagate through the air to your ears.

A radio transmitter acts the same way, except that its amplifier produces energy at a higher frequency than the sound you can hear, and the transducer is an *antenna* that transforms the high-frequency electrical energy into an electromagnetic wave. This wave can propagate through air (or space) for long distances.

For some reason, perhaps because of our familiarity with audio systems or because you can actually hear the results in your ears, it seems easier to grasp the concept of the generation and propagation of acoustic waves than it is to understand the generation and transmission of radio waves.

The audio transmitter analogy can be continued in the receiving direction. A *microphone* is just another transducer that transforms acoustic waves containing speech or music into weak electrical signals that can be amplified and processed. Similarly, a receiving antenna captures weak

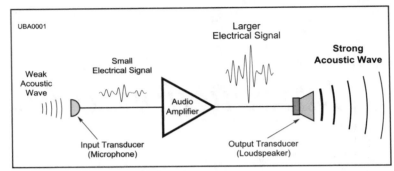

Fig 1-1 — Illustration of transducers changing one form of energy into another.

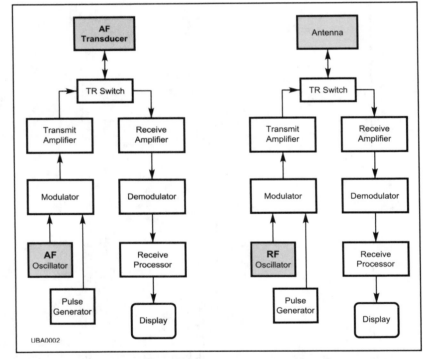

Fig 1-2 — At (A), simplified system diagram of a sonar system. At (B), simplified system diagram of a radar.

electromagnetic waves and transforms them into electrical signals that can be processed in a receiver. **Fig 1-1** shows a system or block diagram of a sound system with transducers (loudspeaker and microphone) at each end.

Many of the phenomena that act upon acoustic waves also occur with electromagnetic waves. It is not an accident that a parabolic reflector radar antenna looks very much like a

parabolic eavesdropping microphone, or that sonar and radar operate in the same fashion. *Sonar* relies on acoustic waves propagating through water to find and reflect back from underwater objects, such as submarines or schools of fish. Similarly, *radar* sends out electromagnetic waves through space, listening for signals reflected back from objects such as aircraft, space vehicles or weather fronts. **Fig 1-2** shows simplified system

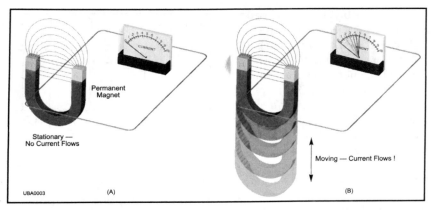

Fig 1-3 — A wire immersed in the magnetic field of a permanent magnet. At A, if everything is static there is no current flow in the wire. At B, if you move the wire up and down within the field, the wire experiences a changing magnetic field and current is induced to flow.

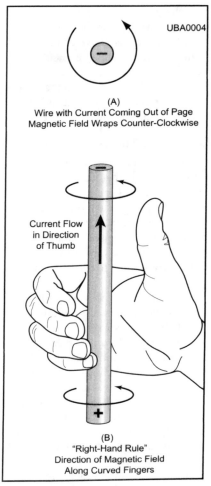

Fig 1-4 — Direction of magnetic field around a wire.

diagrams of radar and sonar systems.

While we're mentally picturing antennas, perhaps an even better analogy than acoustic waves is to compare electromagnetic waves of light to those of radio signals. Light waves propagate through space using the same mechanisms and at the same speed as radio waves. Similar parabolic shapes can reflect and focus both light and radio waves. Light needs a polished mirror as a reflector, such as the mirrored reflector in your flashlight or car headlight. I will sometimes draw on your appreciation of light reflection as I discuss some types of antennas.

What's an Electromagnetic Wave?

An *electromagnetic wave*, as the name implies, consists of a combination of the properties of both electric and magnetic fields.

Static Electric and Magnetic Fields

You are no doubt familiar with magnetism and electricity from everyday experience. One kind of magnetism and electricity is the *static* form. Static electricity is a collection of positive or negative charges that are at rest on a body until discharged by a current flow. This happens to your body when you walk across a rug on a dry day. In climates where the humidity is very low, particularly during the winter, your body can accumulate a charge, perhaps making

your hair stand up. That lasts until you discharge the accumulation by touching a grounded object, such as a screw on a light switch plate, or even the long-suffering dog's nose. Then the charge is dissipated sometimes quite dramatically— with a big spark.

Similarly, a static magnetic field exists around a permanent magnet. You can observe the effect of magnetism on a compass needle or even on a screwdriver.

Static electric and magnetic fields do not result in electromagnetic waves. It is only when a magnetic or electric field is *changing* that you can have electromagnetic radiation. You can visualize the effect by considering some other things that make use of electric and magnetic fields. **Fig 1-3** shows a wire immersed in the magnetic field of a permanent magnet. If everything is static (Fig 1-3A) there is no current flow in the wire. If you change the magnetic field at the wire by moving the wire up and down within the field (Fig 1-3B), the changing magnetic field causes a current to flow. This is how an electric power generator transforms mechanical energy into electrical energy.

You could cause the same effect by keeping the wire in one position and changing the magnetic field around it. You could do this by moving the magnet, but more interestingly you could replace the permanent magnet with an electromagnet and change the current in it. This is just a *transformer*, something with which you

are familiar. When you change the current in the transformer winding, you can move between electric and magnetic fields.

Direction of Fields

Both magnetic and electric fields act in particular directions. A magnetic field acts around the conductor carrying the current. The convention for its direction is that if positive current is flowing in a particular direction, the magnetic field will go around the wire, as shown in **Fig 1-4A**. This can be remembered by calling on the *right-hand rule*. This rule says that if you hold a current carrying wire in your right hand, with the current going in the direction of your thumb, the magnetic field will be in the direction of your curled fingers, as shown in Fig 1-4B.

Electric fields act between areas

with charges of different sense. That is, an area with excess electrons is said to have a negative charge, while an area having a deficiency of electrons has a net positive charge. The electric field goes from the positively charged area to the negatively charged area, as shown in **Fig 1-5**.

Changing Fields in Space

The key to electromagnetic waves is to understand that a changing current in space will cause a changing electric field. The changing electric field will in turn induce a changing magnetic field in space and the two will propagate outward from the source, continuing to the end of the universe.

This surprising fact was predicted by a number of forward-looking scientists, notably James Clerke Maxwell in his *Treatise on Electricity and Magnetism*, published in 1873, written without the benefit of Maxwell actually being able to experience the phenomenon. Maxwell's famous equations defined the relationships between currents and fields in a concise way that remains the basis for all work in the field.

In the following decade, Heinrich Hertz, a professor of physics at Germany's Karlsruhe Polytechnic University, used an electric spark to generate electromagnetic waves. He was able to measure their wavelength and velocity, reinforcing Maxwell's theoretical work.

How Do You Make an Electromagnetic Wave?

As the previous discussion describes, it is conceptually easy to generate an electromagnetic wave. All you need to do is to cause a changing current in a conductor and a wave will propagate outward to the end of the universe. Thus if you were to listen carefully to a radio when you discharge your finger against the wall switch plate, you would have a sudden increase in current and a relatively short-duration electromagnetic wave propagating from your finger.

The resulting wave would not be very strong, nor would it be very useful. It carries little information,

although arguably it could be used to measure the dryness of the carpet, or perhaps the location of your finger, but both are a bit of a stretch. The US Federal Communications Commission (FCC) classifies this kind of radiation as an *unintended emitter*. The generator of an unintended emitter is required not to interfere with licensed users of the radio spectrum.

Electromagnetic waves of this sort are generated every time you turn on a light switch or when the spark plugs in your car fire as the engine turns. In these cases, rather than trying to make antennas, the responsible user attempts to *avoid* the propagation of electromagnetic waves — by shielding the engine compartment or by keeping his hands in his pocket as he walks on dry rugs.

So What if You Do Want to Send Out a Wave?

If you really want to transmit a radio signal, you likely would like to send it on a particular frequency. Note that this is different than the wave resulting from a spark, as used in very early transmitters. Sparks, or other fast changing pulses, contain frequency components across a wide radio spectrum.

A modern transmitter starts with an electrical current that is changing at a particular frequency. Causing that current to flow through a conductor will result in an electromagnetic wave that varies at that frequency. The frequency will be maintained as the wave propagates from the conductor. If the signal is strong enough as it reaches a receiving antenna, a receiver tuned to that frequency will be able to receive the signal and decode any information that was embedded in the signal — You have radio!

This Sounds Simple — What's the Big Deal?

As with many things, a conceptually simple concept only gets complicated when you try to make it happen in a really useful way. There are really only two basic practical issues that arise when dealing with antennas:

- How do you get the antenna to accept the power from the transmitter,

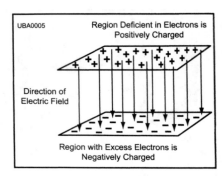

Fig 1-5 — Direction of electric field between two charged areas.

or give it up to a receiver?
- How do you cause the radiation from a transmit antenna to go in the direction you want, or equivalently, how do you get a receive antenna to accept radiation (only) from a particular direction?

There are many peripheral but perhaps equally important issues, such as, "How do you keep the antenna structure from falling down?" In this book, I will focus on the electrical issues and leave the structural issues to mechanical and civil engineers, who have the tools to deal with such matters.

So Why Don't All Wires With Changing Current Radiate?

Well, in fact, they all do. In most cases, however, circuits have a return wire to complete the circuit close to the first wire. Since both these wires carry equal current, they will both radiate equally, but in opposite phase. The net radiation is thus the sum of the radiation from the two wires. But since they are equal and opposite, the net fields are almost zero — except in the region between them. Thus a key requirement for a conductor to be an efficient radiator is that it should not have a return path in close proximity. In this context, *close* means a small fraction of a wavelength at the frequency of the current. A wavelength is the distance a wave travels during the time of one cycle.

Some Concepts You'll Need to be Familiar With

This book is intended to provide a basic understanding of antennas. It would be nice if you could just start here, but you do need to have just a bit of background in some concepts from electronics and radio to make it all come together. These will be introduced from time to time, with the hope that if they are not familiar, you will go back to some of your earlier reference books to review the topic. To start with, you should clearly understand the following terms.

Frequency

This is the rate at which an alternating current waveform or signal goes through a full cycle, from zero voltage to maximum, back to zero, to voltage minimum and then back to zero. The basic unit is hertz (Hz). A signal with a frequency of 1 Hz completes one cycle every second. You will generally be dealing with signals at higher frequencies and the number of Hz will generally be preceded with a modifier indicating that the value is some multiple of Hz. The usual ones you will see are:

- kilohertz (kHz), thousands of Hz
- Megahertz (MHz), millions of Hz, or thousands of kHz
- Gigahertz (GHz), billions of Hz, or thousands of MHz.

Thus a signal with a frequency of 3,600,000 Hz is the same as one of 3600 kHz or 3.6 MHz.

Period

Period is defined as the time it takes for a waveform to make a complete cycle. The basic units are seconds. If a waveform makes F cycles in one second, the time it takes for a single cycle is just 1/F. Thus, in the example above, a 3.6 MHz signal would have a period of 1/3,600,000, which is 0.000000278 seconds. I will use the multiplier of *microseconds* to describe events in millionths of seconds, so the period would be described as 0.278 microseconds, often written as 0.278 µsec, where the Greek letter µ is short for "micro."

Wavelength

Radio signals travel at a finite speed. In free space, or in the Earth's atmosphere, they travel at the same speed as light. Light travels at about 186,000 miles per second, or 300,000,000 meters per second in space. I will generally work in units of meters per second to conform to the practice of most people working in the radio field.

Wavelength is defined as the distance a wave travels during one cycle. In the above example, during the 0.278 µsec that the signal takes to complete a cycle, it will travel 0.000000278 seconds × 300,000,000 meters per second or 83.3 meters. That is the wavelength of a signal at 3.6 MHz. I will generally follow industry practice and use the Greek letter lambda (λ) to indicate a wavelength. Thus, if something were two wavelengths long, you would say 2 λ. For half a wavelength, you would say $\lambda/2$.

A signal can be described equally well by either its frequency or its wavelength. It may be of interest to note that before the then-new FCC started assigning the spectrum by frequency around 1930, the use of wavelength was more common. Many radios of the period had their dials labeled in both frequency and wavelength, since it wasn't clear which term would become the more common system of signal definition. Common usage now is to define particular signals by frequency, while using the approximate wavelength to indicate a band of frequencies assigned to a particular service. Thus an 83.3-meter signal is one of a number of frequencies in the "80 meter" amateur band.

Chapter 2

The Half-Wave Dipole Antenna in Free Space

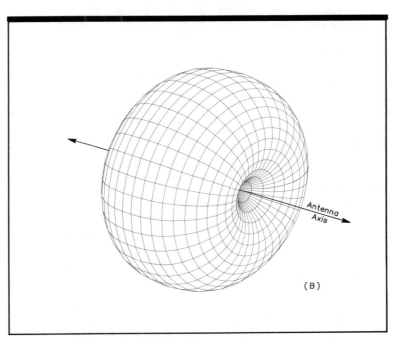

(B)

Radiation pattern of a dipole in free space.

Contents

The Half-Wave Dipole

It's tempting to start out looking at a tiny radiating element consisting of a short conductor with an alternating current flowing in it. Then you could consider all other antenna structures as collections of such *mini-antennas*, each piece with a slightly different location, current magnitude and current phase. The radiating electromagnetic field results from the summation of all the fields radiated by all the mini-antennas. In fact, that's the way many antenna-analysis software programs analyze the performance of complicated antennas.

However, I won't start there because it's hard to imagine just how you can make those mini-antennas work without considering all the interconnecting wires, and these interconnecting wires may well be more significant than the antennas themselves. I will thus start with an examination of an antenna that will actually work and is frequently encountered — the half wave ($\lambda/2$) dipole.[1]

A $\lambda/2$ dipole antenna is formed by having a conductor that is electrically half a wavelength long at the frequency that will be transmitted or received. Here, λ represents the distance the electrical signal would travel down the wire during the

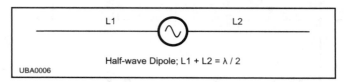

Fig 2-1– Half-wave ($\lambda/2$) dipole.

time of one cycle of the sine wave alternating current (ac) signal from a generator at the operating frequency f_0. I will discuss the antenna first as a transmitting antenna and later discuss receiving applications. This antenna is a good place to start because:

- The $\lambda/2$ dipole is frequently used as an antenna in many real applications.
- Many other antennas are made up of combinations of $\lambda/2$ dipoles in various configurations.
- The $\lambda/2$ dipole is a reference standard for comparison of the performance of other antennas.

The layout of a $\lambda/2$ dipole is shown in **Fig 2-1**. The circle in the center represents a generator that generates a signal at the operating frequency f_0. The total length of the conductor from end-to-end is approximately $\lambda/2$.

The distance of one wavelength in free space is just the speed of light multiplied by the time it takes to make a complete *cycle*. The time for one cycle is the waveform's *period* or $1/f_0$. Thus a wavelength in free space is given by the following formula:

$$\lambda = c/f_0$$

Here c is the speed of light and f_0 is the frequency of the signal that generated the wave. There are many units that can be used; however, it is important that they be consistent. For example, if c is in meters per second, f_0 is in Hertz (cycles per second), then λ will be in meters. For example, for a frequency of 10 MHz, you have

λ = 300,000,000 meters per second $\div$ 10,000,000 Hertz = 30 meters.

Note, for radio work, it is convenient to use a value for c of 300 million meters per second, and f_0 in MHz, which neatly cancels out the six zeros in numerator and denominator.

λ = 300 million meters per second $\div$ 10 MHz = 30 meters.

So How Does a Dipole Work?

The key to understanding how a dipole works is to look at the boundaries or endpoints, where some conditions exert themselves. Physicists would call these *boundary conditions* because they occur at the boundary between regions. In this case you have a boundary between a conducting wire and free space at each end of the dipole. The condition is that there can be no current flow at the open end of a conductor — there's no place for current to go! If the generator is delivering any power to this antenna, and it is, then $\lambda/4$ back from the place where the current is zero, the current must be at its maximum, as shown in **Fig 2-2**.

Similarly, with no current flowing at the end of the antenna, the voltage must be at its maximum. Again, the minimum voltage is found $\lambda/4$ back from the place where it is maximum, as shown in **Fig 2-3**. That would be at the center of the dipole, $\lambda/4$ from each end, just where I've hooked up the generator.

The power the generator puts into the antenna will (you hope) be trans-

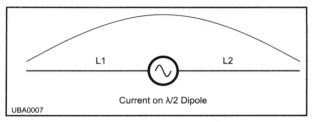

Fig 2-2 – Current along a $\lambda/2$ dipole.

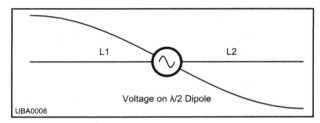

Fig 2-3 – Voltage along a $\lambda/2$ dipole.

formed from an electrical signal into a radiated electromagnetic wave. The antenna power will just be the current at the center multiplied by the voltage at the center, which is, of course, the current and voltage supplied by the generator. Ohm's Law, just as in any other circuit, defines the relationship between the voltage and current. Note that the feed point of the dipole is just

an electrical circuit, with connections going to the antenna from the generator. In *free space*, away from ground (more about free space later) you find that the voltage at that point is just 73 times the current. Thus the generator sees a load that looks like a resistor of 73 Ω = E/I. This value can be rather strenuously derived, but it is much easier to imagine measuring it![2]

Where Does the Power Go?

The amount of power that the generator delivers is the same that it would deliver into a resistor with a resistance of 73 Ω. This forms a convenient boundary between the circuit portion of a system and the radiating-field portion of the system. The generator can be designed to deliver power to a 73 Ω resistor and that power (less any losses) will be transformed into a radiated field by the changing currents flowing on the dipole. This resistance is called the *radiation resistance*.

The radiating field leaves the dipole in a three-dimensional fashion that is a little difficult to describe fully in words. Outgoing electrical and magnetic fields created by the antenna are perpendicular to each other, and each field is also perpendicular to the direction in which a wave travels away from the antenna. The overall field strength in any direction is proportional to the magnitudes of the two field components. This results in a radiating field that leaves with maximum strength in a direction perpendicular to the wire making up the dipole. The field strength is reduced as you move around the dipole towards the direction of the ends, at which point there are neither magnetic nor electric field lines. This is roughly shown in **Fig 2-5**.

A more precise representation of the field strength in each direction as you move from one end half way around the antenna is shown in **Fig 2-6**. Here I have plotted both the relative strength either of the electric or of the magnetic field (since they are proportional), as well as the power, which is proportional to the square of the field strength.

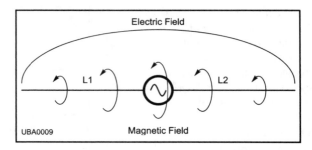

Fig 2-4 Electric and magnetic fields surrounding a λ/2 dipole.

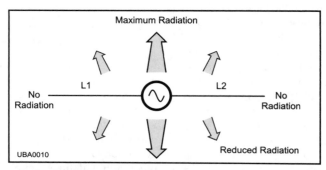

Fig 2-5 – Direction of radiation from a λ/2 dipole.

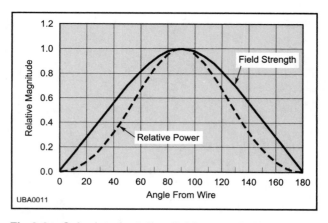

Fig 2-6 – Calculated relative field strength and power.

Antenna Analysis Tools

It is quite possible to manually calculate the field strength from antennas, particularly simple ones such as a dipole. Such calculations require math skills beyond the scope of this book. Fortunately, through the use of computer modeling, you can avoid such efforts. Computer modeling is also of great benefit for the analysis of more complex structures, providing convenient, easy-to-use output formats.

There are a number of such programs available. I will use *EZNEC*, a program written by Roy Lewallen, W7EL, that is easy to use, and which, at this writing, is available in a limited size free demonstration version on his Web site, **www.eznec.com**. A description of the basics of how to use *EZNEC* is provided in Appendix A.

Polar Plots

While a representation as shown in Fig 2-6 can be used to describe the strength of the fields or the power in a particular direction, it is more common to show the information in something called a *polar plot*. This kind of plot represents the intensity in a particular direction by the length of a line, at any angle from the center of the plot to the pattern. **Fig 2-7** is a polar plot of the radiation from a dipole in free space as a function of the azimuth angle with 0° being perpendicular to the antenna. This is one of the outputs from *EZNEC* and you will be using this view extensively throughout the book. The relative power is generally shown in *decibels* (dB), a convenient logarithmic representation that makes it easy to add up system gains and losses. If you are not familiar with decibels, see Appendix B.

Fig 2-8 is a representation of the field strength as a function of elevation angle, taken at the azimuth angle of maximum output. Note that because you are in free space and thus have not considered any ground effects, the elevation pattern is constant. You will see quite a difference when you get "down to Earth" in the next chapter!

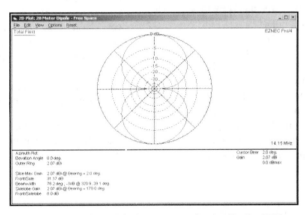

Fig 2-7– Modeled relative power in decibels (dB) as a function of azimuth angle on a polar plot.

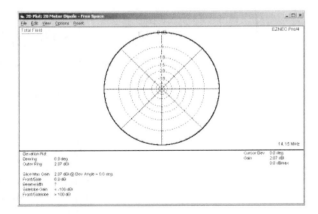

Fig 2-8 – Modeled relative power vs elevation angle.

Which Way is Up?

An antenna in *free space*, such as we have been discussing, is interesting because you don't need to consider the interactions between the antenna and anything else around it. In the extreme case, your antenna is considered to be the only object in the universe. While this might be simple to discuss, it is very hard to actually measure — Where would you stand to make a measurement?

Polarization

In real life many antennas are near, and occasionally even under, the Earth. I will discuss how this affects antenna operation in later chapters. However, it is important at this point to define *antenna orientation*. A dipole antenna constructed near the Earth could be oriented in a number of ways. The extreme cases, and those most often encountered, are with the dipole conductors in a straight line, either parallel to, or perpendicular to, the Earth's surface.

If the dipole is oriented parallel to the Earth, the electric field would also be parallel to the Earth. As seen by an observer on the ground, both the antenna and the electric field would appear to be horizontal. Such an antenna is said to be *horizontally polarized*. Not too surprisingly, a dipole perpendicular to the Earth's surface is said to be *vertically polarized*.

The polarization direction of an antenna is important for a number of reasons:

- Antennas that are horizontally polarized will not receive any signal from a vertically polarized wavefront, and vice-versa.
- Antennas that are horizontally polarized have performance characteristics very different from those vertically polarized when both are near the ground.

Either polarization can be effectively used; however, it is important to understand the differences so that you can make good use of their distinct properties. Note also that there is nothing that says an antenna can't be oriented in-between horizontal and vertical. Such an antenna is said to have *skew polarization*. It can be considered a combination of horizontal and vertical polarization, with part of its power associated with each, depending on how far it is tilted.

It is also possible to have an antenna that generates a wavefront that shifts in polarization as it leaves the antenna, continually changing in space. This is called *circular polarization,* and I will discuss it, as well as skew polarization, in applications later in this book.

Notes

[1] See Chapter 1.

[2] J. Kraus, W8JK, *Antennas*, McGraw Hill Book Company, New York, 1950, pp 143-146.

Review Questions

2.1. Describe circumstances for which a "free space" dipole model can represent a real antenna.

2.2. Calculate the approximate length of $\lambda/2$ dipoles for 0.1, 1, 10, 100 and 1000 MHz.

2.3. Discuss applications for which a horizontally polarized dipole might be most appropriate. Repeat for a dipole with vertical polarization.

2.4. Consider an amplifier with 20 dB of gain, matching networks at input and output each with a loss of 1 dB and an antenna with a gain of 5 dB. What is the total system gain? (See Appendix B, if needed.)

2.5. If an input signal of 0 dBm is applied to that system, what is the output radiated power in dBm, mW?

Chapter 3

The Field From a Dipole Near the Earth

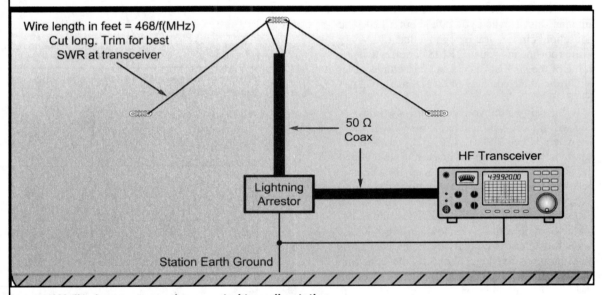

Wire length in feet = 468/f(MHz)
Cut long. Trim for best
SWR at transceiver

50 Ω
Coax

HF Transceiver

43992000

Lightning
Arrestor

Station Earth Ground

Inverted V dipole over ground connected to radio station.

Contents

I have discussed the way a dipole works if far removed from the Earth. While this is a useful place to begin, and may be directly applicable for space communications, I now want to discuss the very real case of antennas near the Earth. The Earth has two major effects on antenna performance and behavior:

- Reflections from the Earth's surface interact with radiation leaving directly from the antenna, resulting in a change to the direction at which that radiation seems to leave the antenna.
- Proximity to the Earth may change the electrical parameters of the antenna, such as its feed-point *impedance*, as I'll get into later.

You must first grasp the fundamental concept of the phase of a wavefront.

Phase of an Electromagnetic Wave

In the last chapter, when I spoke of the strength of an electromagnetic wave, or its parts — the electric and magnetic fields — I was talking about the magnitude of an ac sine wave at the frequency of the generator driving the antenna. Just as when you talk about a 120 V ac household circuit, you recognize that 120 V is the root-means-square (RMS) or effective value of the sine wave. The actual voltage varies with time at a frequency of 60 Hz, as shown in **Fig 3-1**.

If you sample the voltage at various times, you will measure an instantaneous voltage anywhere from −170 to +170 V. If you were to combine two such signals in a circuit that added the voltages, the result could range anywhere between −340 to +340 V, depending on the relative *phase* of the two signals. If you were to add two signals of the same phase, you would end up with twice the voltage, as shown in **Fig 3-2**.

On the other extreme, if one signal were at its maximum positive level at exactly the same time as the other is at its maximum negative level, the resulting net voltage would add up to 0 V. In other words, the signals would *cancel* each other. **Fig 3-3** shows a signal whose phase is exactly out-of-phase with the signal in Fig 3-1.

Electromagnetic waves travel from an origin to a destination via multiple paths, especially through the ionosphere, and as a consequence they exhibit all sorts of variations. Depending on the relative phase of the signals at the reception point, they could add, or they could subtract from each other. Consider the case of only two signals arriving at a receiver with exactly the same strength. They can combine to produce a signal that is somewhere between twice the level of each signal by itself, down to a level of zero, where they completely cancel each other. *Fading* and signal enhancement can occur for many different reasons.

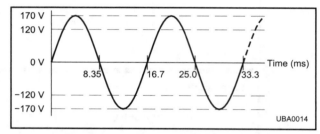

Fig 3-1 – Voltage of a 120 V ac sine-wave signal as a function of time.

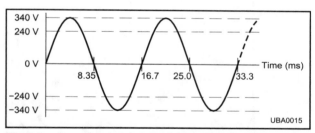

Fig 3-2 – Resultant voltage of two signals of Fig 3-1 adding together versus time.

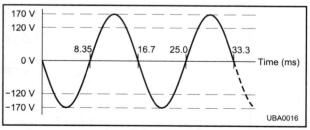

Fig 3-3 – Voltage of a second 120 V 60-Hz ac signal opposite in phase to Fig 3-1.

What is the Effect of Ground Reflections?

A signal transmitted in any direction from an antenna near the Earth will have a direct path, and it will also have a path that results from a reflection from the surface of the Earth. Consider a simple case to start. Assume that you are an observer located some distance from the antenna such that the Earth between you and the antenna looks like it is flat. Assume that the antenna is mounted over perfectly conducting ground. This is depicted in **Fig 3-4**.

For any elevation or *takeoff* angle there is a direct path from the antenna to the observer, and there is also a reflected path. These two paths have different lengths, with the reflected path always longer than the direct path. The difference in length will depend on how high the antenna is above the ground and the elevation angle. Note that this *flat Earth* model falls apart for the case of zero elevation because there can be no reflection, but ignore that extreme case for the time being.

As the downward wave strikes the Earth, the combination of the incident wave and reflected wave cannot create an electric field at the surface of the ground since the fields can't exist in a perfectly conducting ground medium — they are, in essence, shorted out by the ground. For a horizontally polarized antenna in Fig 3-4 the reflected wave therefore must be out-of-phase with the incident wave. I have indicated the polarization of a horizontally polarized antenna with a "+" sign. This represents the tail of an arrow (or vector) heading into the paper.

Interestingly, and importantly, a vertically polarized wave must have the reflected and incident waves in-phase with each other because opposite ends of the field are at the Earth's surface when they are in-phase.

Note that for any takeoff angle and height, you could calculate the difference in path length using plane geometry and trigonometry. By knowing the difference in path length, the signal frequency and the speed of propagation (which is the same as the speed of light in air), you could easily compute the phase difference due to the different path and thus the resultant amplitude. Fortunately, you can also use antenna-modeling tools to determine the same thing.

You also can imagine that the reflected wave comes from another antenna that is located under the earth the same distance that the real antenna is above the earth. This is called an *image antenna* and isn't real, but the path length difference is a bit easier to visualize and calculate. Again, the two antennas are in-phase for the case of vertically polarized antennas and out-of-phase for horizontal ones. The configuration is shown in **Fig 3-5** for horizontal antennas.

Note the point of the arrowhead on the image antenna, indicting opposite polarity from the real antenna. The vertical antenna configuration is shown in **Fig 3-6**. The height is at the center of the antenna. Note that the antenna and its image have the same polarity for the vertical case.

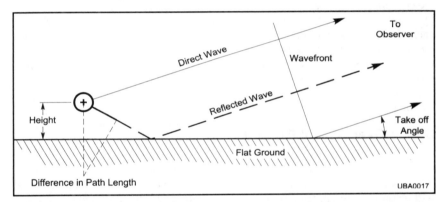

Fig 3-4 – Illustration of additional path length (and thus phase delay) of reflected wave compared to direct space wave

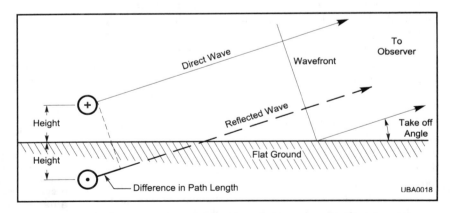

Fig 3-5 — Image antenna concept for visualizing ground reflection. Note opposite polarity (180° phase shift) of image for horizontal polarization.

How Do The Numbers Add Up?

Knowing the phase of the reflected wave and the height of the antenna, you can thus determine the resultant phase of the direct and reflected signals as a function of height. This just means that you need to know the difference in path length in terms of wavelengths. For example, if the polarization is vertical, and the difference in path length is an odd multiple of a half wavelength, the signals will be out-of-phase and cancel at that angle. At other elevation angles, the difference may be an even number of half wavelengths and the signals will be the same phase and will add together. For horizontally polarized waves, it is just the reverse. Intermediate angles will have values in between these extremes.

You can determine the intensity of the combination of the two antennas by merely adding up the signals for each elevation and azimuth angle. Alternately, you could take advantage of the capabilities of an antenna-analysis program to do so, such as *EZNEC*, which I've mentioned previously in Chapter 2.

Dipole Over Typical Ground

As you would expect based on the earlier discussion, the elevation pattern of an antenna near the Earth will be quite different from one far removed from the Earth. This is all due to reflections from the Earth. If you move the antenna with the modeled uniform elevation shown earlier in Fig 2-8 from outer space down to ½ wavelength above ground (about 50 feet at this frequency), you will get the pattern shown in **Fig 3-7**, which compares the elevation pattern for a dipole ½ wave above both *perfect ground* and typical soil. First, consider the reflected wave. Remember for a horizontal antenna at this height the reflection is out-of-phase to start with. So ½ wavelength (180°) off ground gives a phase reversal at the reflection (another 180°) and one more ½ wavelength (180°) up towards the antenna result in an out-of-phase signal that cancels the upward going wave. Note also that the wave along

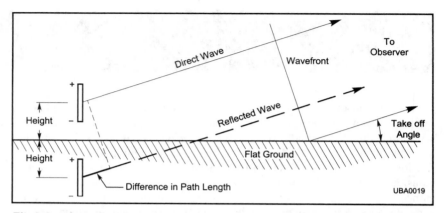

Fig 3-6 — Image antenna concept for visualizing ground reflection. Note same polarity (0° phase shift) of image for vertical polarization.

the horizon cancels with the out-of-phase reflected wave, resulting in no radiation at 0° elevation.

Where Does the Power Go?

Nothing in the process I just described heats up or otherwise absorbs power, so the total power is redistributed to the areas that aren't reduced by the reflected signal. The areas with the most signal have a significantly stronger signal than they had in the free-space case, because the other areas have a significantly weaker signal. This effect is referred to as *ground reflection gain*. It isn't a real gain, such as you might get from an amplifier, but is more of a redistribution. On the other hand, if you want the signal to go where the signal combined with its ground reflection goes, it seems just like an amplifier to a distant receiver.

This redistribution happened to a certain extent with the dipole in free space described in Chapter 2. Note that because the radiation does not occur from the ends of the dipole, the main beam is about 2 decibels (dB) stronger than if it were a completely uniform *isotropic radiator* (a theoretical antenna that radiates equally in all directions). By tending to cancel the upward and horizontal signals, the maximum signal in the main beam is about 5.5 dB stronger than for

the free-space case. This can be a real advantage if that's where you want the energy to go!

What About a Better Ground?

The model used for Fig 3-7 was an attempt to model typical dirt. For those who like the details, it assumed ground with a conductivity of 0.005 siemens/meter and a dielectric constant of 13. *EZNEC* allows you to enter the exact ground parameters for your location, or you can choose a perfect ground model. For perfect ground, imagine a few acres of gold foil under the antenna, heading off in all directions. The results are also shown in Fig 3-7. Note that the general shapes are similar. However, the upward cancellation is complete over perfect ground since the wave is completely reflected from a perfect conductor. The resulting ground reflection gain is a bit higher as a

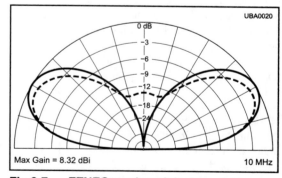

Fig 3-7 — *EZNEC* overlay of the broadside elevation pattern of a horizontal dipole antenna mounted a half wavelength over real ground (dashed line) compared to that same antenna over perfect ground (solid).

Table 1

Lowest Elevation Peak and Ground Reflection Gain at Various Elevation Angles for Dipole Over Real Ground

Height Above Ground	Center of Peak	Gain at 10°(dBi)	Gain at 20°(dBi)	Gain at 30°(dBi)	Gain at 90°(dBi)
¼ Wavelength	90°	−3.7	1.5	3.9	5.7
½ Wavelength	28°	2.55	6.71	7.38	−5.5
1 Wavelength	14°	6.86	5.8	−10.6	−5.6
2 Wavelengths	7°	5.9	6.9	−9.1	−4.2

consequence.

While perfect ground may be hard to come by, *saltwater* provides a close approximation. It is also possible to simulate almost-perfect ground over a region with a large expanse of bonded wire mesh, or similar structures.

What Happens at Different Heights?

If you examine again the geometry in Fig 3-4 or Fig 3-5, it's clear that the elevation pattern of a horizontal antenna is very dependent upon the height above ground. If the antenna is much lower than the ½ wavelength you have been looking at, a horizontal antenna will not have the upward direction energy cancelled, with the result that most of the energy heads upward. This is shown for the case of a λ/4 high dipole in **Fig 3-8**, which overlays the responses for three horizontal dipoles — ¼, 1 and 2 wave-

lengths high. Later in this book when I discuss how signals get from place to place, you'll discover that a low antenna can work well for medium distance communications.

As the height over ground increases, the patterns for a horizontally polarized antenna tend to get more complex, and you get an increased number of elevation angles with *nulls*. This is shown clearly in the case of a horizontal antenna a full wave above ground and for one that is two wavelengths above ground in Fig 3-8. Note that as the antenna height increases, the first radiation peak moves down to lower angles and each peak covers a narrow range of elevation before the next null. This results in gaps in elevation-angle coverage. A summary of the signal intensity for each case is shown in **Table 1**.

I haven't mentioned the azimuth pattern in a while, largely because

not much happens as the antenna height is changed. Above the very lowest heights, it stays about the same. Compare **Fig 3-9**, a plot of the azimuth pattern of a horizontal dipole mounted 2 wavelengths above ground with that of the horizontal dipole in free space (Fig 2-7) and note the similarity.

How About Vertically Polarized Antennas?

So far I have been discussing horizontally polarized dipoles. I could have just as well started with antennas with vertical polarization near the ground. As noted in Fig 3-6, the geometry is the same, but the big difference is that the signal from the image is in the same phase as that from the antenna. This means that the signals add towards the horizon for perfect ground rather than having a null at 0° elevation. The elevation

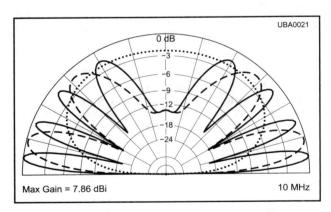

Fig 3-8 — *EZNEC* overlay of broadside elevation patterns of a horizontal dipole mounted at three heights over real ground: 2 λ = solid line, 1 λ = dashed line, λ/4 = dotted line.

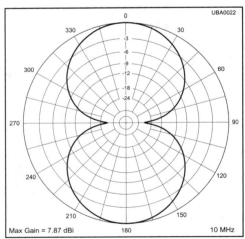

Fig 3-9 — *EZNEC* azimuth plot for a horizontal dipole mounted 2 λ above typical ground at the peak of the first elevation lobe (7°).

pattern of a ½ wave dipole whose bottom is 1 foot over perfect ground is shown in **Fig 3-10**, along with a plot of the same antenna mounted 1 foot over typical soil. Note that unlike the horizontal dipole, either vertical dipole radiates equally well at all azimuth angles, often an advantage for some types of systems such as broadcast or mobile radio. You can see very clearly in Fig 3-10 that even typical soil has a big effect on the level of signal launched by a vertical antenna. This is due to the losses incurred when the signal is reflected from lossy soil.

The effects of ground reflections are also apparent for vertical antennas as they are elevated. **Fig 3-11** shows a comparison when the bottom of a vertical dipole is elevated one and two wavelengths above typical soil. You can see that elevating a vertical dipole well above lossy soil has a strong effect on the strength of signals launched from that antenna.

Fig 3-12 compares the elevation response for a vertical dipole whose bottom is 2 wavelengths high and a horizontal dipole that is 2 wavelengths high. This is again over typical soil.

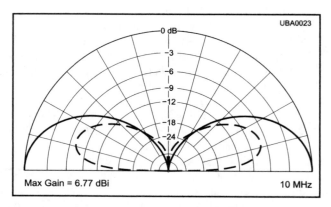

Fig 3-10 — *EZNEC* overlay of elevation plots for a vertical dipole antenna whose bottom is mounted 1 foot above typical ground, compared to the same dipole mounted over perfect ground. You can see that reflection losses in typical soil can be high.

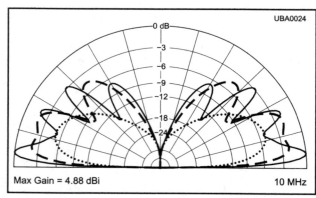

Fig 3-11 — *EZNEC* overlay of elevation plots for a vertical dipole antenna whose bottom is mounted 1 foot above typical ground (dotted line), compared to the same dipole mounted 1 λ above typical ground (dashed line) and the same dipole mounted 2 λ over typical ground (solid line). The loss in the ground directly under the antenna (sometimes called "heating up the worms") can be substantial when the bottom of the antenna is close to the soil.

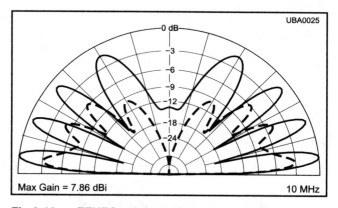

Fig 3-12 — *EZNEC* overlays of elevation plot for a vertical dipole antenna whose bottom end is mounted 2 λ above real ground (dashed line), compared with a horizontal dipole mounted 2 λ high (solid line).

Chapter Summary

In this chapter I have examined the effects of moving a simple antenna from outer space to a practical location near the ground. You saw that reflections from the ground result in signals that can add or subtract from the antenna's main wave, depending on the relative phase of the two signals. This effect can prove to be beneficial or detrimental, depending on how the system will be used. In any event, it is easily predictable through modeling and you can decide, as a system designer, how you want everything to come together.

The Earth is but one reflection mechanism with which you will have to deal. Almost any object that a signal encounters will change that signal to a certain extent. Sometimes, as in radar, a reflected signal is the reason for the system. In other circumstances, it can be either helpful or cause problems. I will explore circumstances of each type in later chapters.

Review Questions

3.1. Calculate the actual height of an antenna $\lambda/2$ above the ground for frequencies of 1, 10 and 100 MHz.

3.2. Compare Figs 3-7 and 3-8 and consider why less-than-perfect ground may still be fine for horizontally polarized antennas. Under what conditions would perfect ground help you?

3.3. Repeat question 3.2. for vertically polarized antennas. Compare Figs 3-10 and 3-12 to get the idea. Why might you want to take extra care to simulate a perfect ground for a low vertical dipole?

3.4. Based on their azimuth and elevation patterns, can you think of applications that would be best suited for vertical antennas? How about horizontal antennas?

Chapter 4

The Impedance of an Antenna

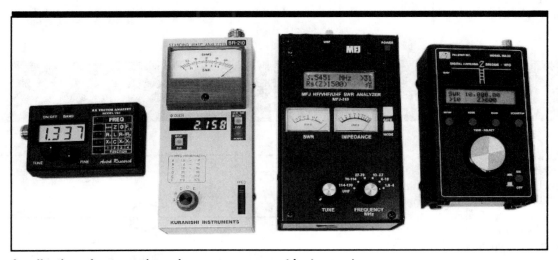

A collection of antenna impedance measurement instruments.

Contents

So far I have discussed the half-wave dipole in some of its different forms. I have always assumed that there was a transmitter (or receiver) at the center of it. I haven't yet discussed how you connect the transmitter to the antenna, or whether it's convenient to physically locate the transmitter at the center of the antenna. There are really two issues here:

- How does the feed-point input impedance of the antenna compare to the load impedance needed by the transmitter? This chapter will discuss the nature of antenna feed-point impedances.

- The second issue is whether or not you can (or would want to) physically locate the transmitter at the

center of the antenna. This is sometimes the case, and I could describe some examples, but it is generally reserved for situations in which the antennas are relatively large and the transmitters are relatively small, such as phased-array radars. In other cases, you generally interconnect the transmitter and antenna with a *transmission line*.

A transmission line is a type of cable designed for the purpose. They are available in a number of configurations, all with the property that if they are connected to an antenna (or any load for that matter) that has an impedance equal to the *characteristic impedance* of the transmission line, a value determined by the physical properties of the line, the other end of

the line will see the same impedance. Thus if we *match* the impedance of the antenna to the transmission line, we can locate the transmitter any distance from the antenna and have it act almost as if the transmitter were at the center of the antenna.

We will discuss transmission lines, as well as matching, in the next chapter. The place to start is with the impedance of the antenna and the knowledge that you don't really have to locate the transmitter at the middle of it. Anyone who has seen a photo of the transmitter room at a TV or radio station or especially the Voice of America will immediately appreciate the desirability of being able to separate the transmitter from the antenna!

Antenna Impedance

As with any circuit element, the impedance an antenna presents to its source can be defined by the current that flows when a voltage is applied to it. There are, after all, no footnotes to Ohm's law that repeal it for the case of an antenna. Thus for any possible connection point to an antenna, if we know the levels of current and voltage, Ohm's law will reveal the impedance at that point.

Note that I could talk about a receiver connected to the antenna rather than a transmitter. Here, I will discuss transmitting antennas, with the understanding that receiving antennas have the same impedance characteristics.

The Impedance of a Center-Fed Dipole

In Chapter 2, I discussed the current and voltage distribution along the length of a λ/2 dipole. I elected to feed it in the center since that's where the voltage was at a minimum and the current at a maximum. The drawings of current and voltage are reproduced as **Fig 4-1**. The ratio of voltage to current, the impedance, will vary as you change any of the key dipole

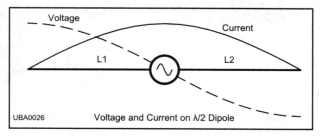

Fig 4-1 — Voltage and current distribution along the length of a resonant half-wave dipole.

parameters. If you change the length through values around λ/2, the ratio will go through a point at which the ratio is resistive. This is also described as the *resonant* point— that is, there is no reactive component and thus the impedance is entirely resistive. This special length is referred to as the resonant half-wave dipole length. Even though it is special, it is very useful and frequently encountered.

Impedance of a Dipole in Free Space

I will discuss some of the factors that cause the impedance of a λ/2

dipole to differ; however, as a starting point, I'll consider the case of a thin dipole in free space. At resonance it will have an impedance of around 72 Ω. If you make the antenna just a bit shorter (or change to a slightly lower frequency), it will look like a resistance in series with a small amount of capacitance. If you make it just a bit longer, it will look like a resistance in series with a small inductance.

A key parameter in determining both the impedance of a resonant dipole and how the impedance changes with frequency is the ratio of length-to-diameter. Using a 10-MHz

Table 4-1

Impedance of Nominal λ/2 Dipole in Free Space (Ω, – indicates capacitive reactance)

Frequency MHz	L/D = 10,000 (47.81 foot length) Impedance		L/D = 1000 (47.31 foot length) Impedance		L/D = 100 (45.98 foot length) Impedance	
	R	X	R	X	R	X
9.9	70.0	−16.0	69.7	−11.7	69.6	−7.3
9.95	71.0	−8.1	70.8	−5.9	70.1	−3.7
10.00	72.1	0	72.1	0	72.0	0
10.05	72.2	+8.0	73.0	+5.8	73.2	+3.5
10.1	74.3	+16.1	74.1	+11.8	74.4	+7.2

dipole as an example, **Table 4-1** provides the results from an *EZNEC* simulation of an ideal loss-free dipole at three different length-to-diameter ratios. The first, 10,000:1, is fairly typical of a dipole made from wire. The second would correspond to an antenna constructed from fairly thick wire, while the third would represent an antenna made from 5-inch tubing (or more likely at this frequency, a cage of wires). As frequencies go up and down, the typical length-to-diameter ratios change due to material availability, but all can be encountered in the real world.

There are a few points that should be observed as you look at Table 4-1:

1. As the conductor diameter increases, the length of the resonant dipole decreases. Note that in free space, λ/2 at 10 MHz is 49.2 feet, compared to 47.81 feet for the wire case (about 97%) down to 45.98 feet for the very thick dipole (about 93.5%).

2. As the conductor diameter increases, the change in impedance with frequency decreases. This will be an important consideration when we talk about wideband antennas later in the book.

3. If the source is designed to feed a resistive load, it is relatively simple to provide a match to it even if the antenna has a reactive component. The circuit is shown in **Fig 4-2**. For example, if we want to operate a 10-MHz, L/D = 1,000:1 antenna on 10.1 MHz, the inductive reactance component is +11.8 Ω. By inserting a capacitor with a capacitive reactance of −11.8 Ω at the antenna

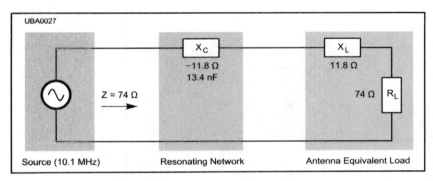

Fig 4-2 — Simplified diagram of an antenna as a load, with a resonating network to provide a resistive load to the source.

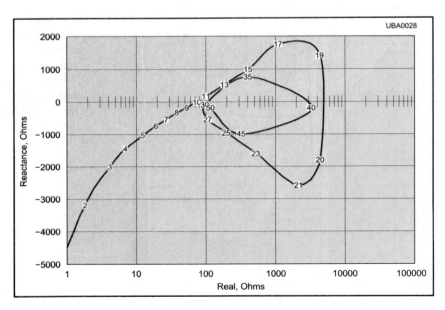

Fig 4-3 — Impedance of 10 MHz, λ/2 dipole with length-to-diameter ratio (L/D) of 10,000:1 from 1 to 50 MHz. The X axis is real (resistive) component, while the Y axis is reactive. Positive values indicate inductive reactance; negative values indicate capacitive reactance. Key frequencies are shown.

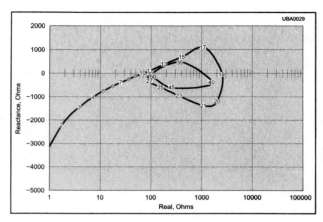

Fig 4-4 —Same as Fig 4-3 except L/D is 1,000:1.

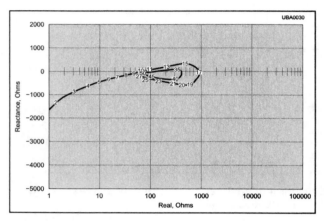

Fig 4-5 —Same as Fig 4-3 except L/D is 100:1.

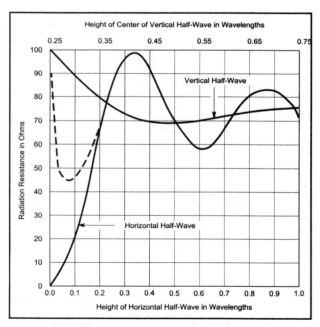

Fig 4-6 — Impedance of a resonant thin dipole as a function of height above ground, or other reflecting surface. The solid line represents a perfectly conducting surface, while the dashed line represents "real" ground.

(see Fig 4-2), you will present the source with a resistive load of 74.4 Ω, almost the same as if we shortened the antenna to make it resonant.

Figs 4-3 through **4-5** show the change in impedance over a wide frequency range for the length-to-diameter ranges of Table 4-1. The differences are striking.

Impedance of a Dipole near Earth

Reflections from the ground couple to a dipole, much in the way a load on one winding of a transformer couples to another. Thus the impedance of a dipole will be different at different heights above ground, depending on the magnitude and phase of the reflection — functions of ground characteristics and the height above ground. **Fig 4-6** shows the impedance of resonant horizontal and vertical dipoles at different heights above ground. Note that other conductive surfaces will have a similar effect on impedance. Coupled antenna elements will also, but I'll reserve that discussion for later.

Review Questions

4.1 Why might we care what the impedance of an antenna is?
4.2 What principle accounts for the difference in the effect of ground between horizontal and vertical antennas?
4.3 What might be the effect of connecting an antenna and a transmitter with different impedances?
4.4 Describe advantages and disadvantages of thin and thick antennas as shown in Figs 4-3 through 4-5.

Chapter 5

Transmission Lines

Transmission lines come in many forms serving many applications.

Contents

As I have mentioned previously, frequently the antenna and radio are not located in the same place. There are some notable exceptions, particularly in portable handheld systems and various microwave communications and radar systems. But in most other cases, optimum performance requires the transmitter and receiver to be at some distance from the antenna. (There may also be a matter of combat survival, especially if your enemy is equipped with anti-radiation weaponry designed to home in on a signal.)

The component that makes the interconnection is called a *transmission line*. Transmission lines are used in places besides radio systems — for example, power-distribution lines are a kind of transmission line, as are telephone wires and cable TV connections.

In addition to just transporting signals, transmission lines have some important properties that you will need to understand to allow you to make proper use of them. This section briefly discusses these key parameters.

Characteristic Impedance

A transmission line generally is composed of two conductors, either parallel wires, such as we see on power transmission poles, or one wire surrounding the other, as in coaxial cable TV wire. The two configurations are shown in **Fig 5-1**. Either type has a certain inductance and capacitance per unit length and can be modeled as shown in **Fig 5-2**, with the values determined by the physical dimensions of the conductors and the properties of the insulating material between the conductors.

If a voltage or signal is applied to such a network, there will be an initial current flow independent of whatever is on the far end of the line, but based only on the L and C values. The initial current will be the result of the source charging the shunt capacitors through the series inductors and will be the same as if the source were connected to a resistor whose value is equal to the square root of L/C. If the far end of the line is terminated in a resistive load of the same value, all the power sent down the line will be delivered to the load. This is called a *matched* condition. The impedance determined in this way is called the *characteristic impedance* of the transmission line and is perhaps the most important parameter associated with a transmission line.

Common coaxial transmission lines have characteristic impedances (referred to as Z_0) between 35 and 100 Ω, while balanced lines are found in the range of 70 to 600 Ω. What this means to us as radio people is that if we have an antenna that has an impedance of 50 Ω and a radio transmitter designed to drive a 50 Ω load, we can connect the two with any length of the appropriate 50 Ω coaxial cable and the transmitter will think it is right next to the antenna. The antenna will receive most (see next section) of the transmitted power and all is well with the world!

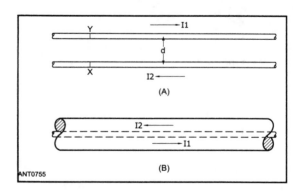

Fig 5-1 – Parallel wire (A) and coaxial (B) transmission lines.

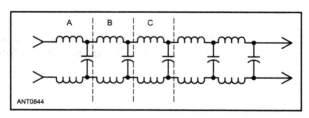

Fig 5-2 – Lumped constant equivalent of an ideal transmission line.

Attenuation

The ideal transmission line model shown in Fig 5-2 passes all input power to a matched load at the output. A real transmission line, however, has loss resistance associated with the wire conductors and loses some signal due to the lossy nature of the insulating material. As transmission lines are made of larger conductors, the resistance is reduced and as the dielectric material gets closer to low-loss air, the losses are reduced. The *skin effect* causes currents to travel nearer to the surface of the conductors at higher frequencies, and the effective loss thus increases as the frequency is increased.

Fig 5-3 provides some real-world examples of the losses as a function of frequency for the most common types of transmission line. Note that the loss increases linearly with length and the values are for a length of 100 feet. Note also that the losses shown are for transmission lines feeding loads matched to their Z_0. As will be discussed shortly, losses can increase significantly if the line is not matched.

The "open-wire" line shown consists of two parallel wires with air dielectric and spacers, typically resulting in a Z_0 of 600 Ω. While the losses of such a line are low, they only work well if spaced from metal objects and not coiled up. While co-axial cables have higher loss, almost the entire signal is kept within the outer conductor. Coaxial cable can be run inside conduit, coiled up, placed next to other wires and is therefore much more convenient to work with. Sometimes a long straight run of open-wire line will be transformed to 50 Ω at the ends with coaxial cable used at the antenna and radio ends to take advantage of the benefits of both.

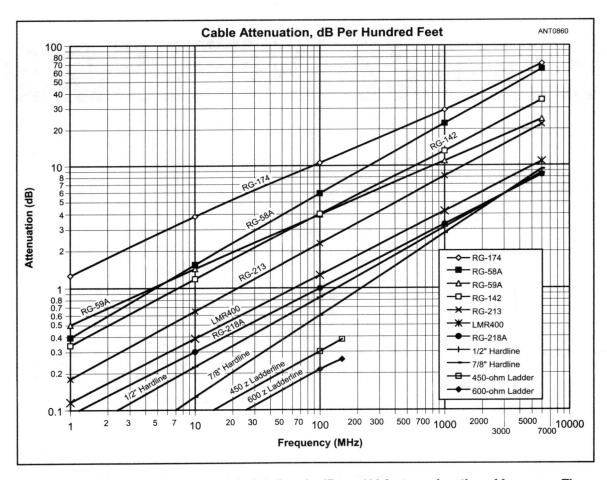

Fig 5-3 — Loss of some typical transmission lines in dB per 100 feet as a function of frequency. The RG-58 transmission lines are 50 Ω polyethylene insulated coaxial cable slightly less than ¼ inch in diameter. The RG-8 through RG-216 are 50 and 70 Ω polyethylene insulated transmission lines with a diameter somewhat less than ½ inch. The "hardline" types have a foam dielectric very near to air.

Propagation Velocity

Signals in air dielectric transmission lines propagate at almost the speed of light. Other dielectric materials cause the signals in transmission lines to slow down, just as you can observe with light rays traveling through water. In many cases, this is not a matter of concern, since you usually only care that the signals get out the other end; however, there are some exceptions.

The velocity is reduced by a factor of one over the square root of the relative dielectric constant. Some cable specifications provide the relative velocity as a fraction of the speed of light. If not, and you know the material, most engineering handbooks include tables of properties of materials. For example, polyethylene is a common cable insulating material and has a relative dielectric constant of 2.26. The square root of 2.25 is 1.5, so the propagation velocity in polyethylene insulated coaxial cable is $3/1.5 \times 10^8$ or 2×10^8 m/sec. The velocity of light in air is 3×10^8 m/sec.

Some applications actually use coaxial cables to provide delayed signals in pulse applications. Having a way to accurately predict the delay just by knowing the cable characteristics and measuring the length of the cable can save a lot of lab time. In the radio world, we will talk about driving antenna elements in a particular phase relationship to obtain a desired antenna pattern. If a transmission line is used to provide the two signals of different phase, we need to know how fast the signal propagates in order to determine the required line length. As we will discuss in the next section, transmission lines of particular electrical lengths can be used as impedance transformers. Unless we know the propagation velocity we can't determine the proper length.

Lines With Unmatched Terminations

In our discussions so far, we have been talking about transmission lines feeding terminations matched to their characteristic impedance. Other than that case, the voltage-current relationship at the load will reflect the impedance of the load — not the characteristic impedance. Along the line the voltage and current will vary with distance, providing a load to the transmitter end that is generally neither that of the far end Z_L, nor the Z_0 of the transmission line. The transmitter load can be calculated knowing the Z_L, the Z_0 and the electrical length of the line as discussed in the following sidebar.

The ratio of maximum voltage on the line to minimum voltage on the line is called the *standing wave ratio* or SWR. A matched line has an SWR of 1:1. A 50 Ω line terminated with a 25 or 100 Ω load will have an SWR of 2:1. There is a whole family of complex impedances that will also have a 2:1 SWR. By the way, the computation of SWR is easier with resistive loads.

There are some interesting special cases with a mismatched line. The load impedance, resistive or complex repeats every λ/2, for example. The impedance goes to the opposite extreme in odd multiples of a λ/4. For example, our 25 Ω load would get transformed to 100 Ω in λ/4 or 3/4λ transmission line sections and vice versa. This effect can be used to our advantage if we wish to transform impedances at a specific frequency.

A generally less desirable effect of mismatched lines is that the losses increase. This is easy to see, if voltages and currents are higher, we might expect losses to increase as well. **Fig 5-4** provides the additional loss for a mismatched line that needs to be added to the matched loss in Fig 5-3. As is evident, the combination of matched loss and high SWR results in dramatic increases in loss. This is why antenna designs that don't use matched transmission lines often use air-dielectric lines, which have inherently less loss under matched conditions to start off with.

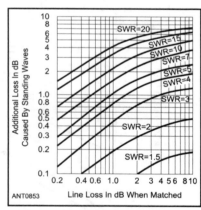

Fig 5-4 – Additional loss of a transmission line when mismatched. This loss needs to be added to the loss in Fig 5-3 for mismatched lines.

Notes

[1] *TLW* is supplied with *The ARRL Antenna Book*, 21st Edition, Available from your ARRL dealer or the ARRL Bookstore, ARRL order no. 9876. Telephone 860-594-0355, or toll-free in the US 888-277-5289; **www.arrl.org/shop/**; **pubsales@arrl.org**.

Determining the Input Impedance of an Unmatched Transmission Line

The input impedance of a transmission line of any length with any terminating impedance can be determined in a number of different ways. The most straightforward way is through direct calculation. Unfortunately, this is also perhaps the most time consuming and perhaps most error prone method, at least until you have it set up on a spreadsheet or other program. The input impedance can be found as follows:

$$Z_{in} = Z_0 \frac{Z_L \cosh(\gamma\ell) + Z_0 \sinh(\gamma\ell)}{Z_L \sinh(\gamma\ell) + Z_0 \cosh(\gamma\ell)}$$

where
Z_{in} = complex impedance at input to line in Ω
Z_L = complex load impedance at end of line = $R_a \pm j X_a$
Z_0 = complex characteristic impedance of the line
ℓ = physical length of the line
γ = complex loss coefficient = $\alpha + j\beta$
α = matched-line loss attenuation constant in nepers/meter. (Note one neper = 8.688 dB. Cables in US are generally specified in dB/100 feet.)
β = phase constant of line in radians/unit length. Note that 2 π radians = one wavelength.

Alternately,
$\beta = 2\pi/(VF \times 983.6/f \text{ (MHz)})$ for ℓ in feet

If the line is lossless (or is short enough that losses are not significant), the expression reduces to the somewhat simpler:

$$Z_{in} = Z_0 \frac{Z_L + jZ_0 \tan(\beta\ell)}{Z_0 + jZ_L \tan(\beta\ell)}$$

Note that most scientific calculators, including the *Windows* scientific calculator and Microsoft *Excel*, can be used to evaluate the hyperbolic trig functions, even if you're not comfortable with them. If you set up (and save) an *Excel* worksheet with the formula above, you will be ready to quickly determine the input impedance any time you have all the input parameters. As noted, be careful with the units.

My favorite way to determine the input impedance, as well as the line loss, standing wave ratio (SWR) is using *TLW* (*Transmission Line for Windows*) software that comes with *The ARRL Antenna Book*.[1] The main screen is shown in **Fig 5-A** performing an analysis of

the condition of the L/D of 10,000:1 case from Table 4-1 at 10.1 MHz. Note the antenna input impedance (74.3 + j 16.1) is inserted in the LOAD box. The output impedance through 100 feet of RG-58A 50 Ω coax is provided at the bottom in both rectangular (69.02 – j 6.23) and polar coordinates (69.53 at an angle of –5.15°), along with the SWR at line input (1.40) and output (1.62) as well as line loss (1.661 dB) — both for the matched case (1.555 dB) and the additional loss due to mismatch (0.106 dB). That's about everything I could think to ask, except which pile in the basement actually holds my RG-58A!

A third way to evaluate the input impedance is through a graphical method. A *Smith chart*, see **Fig 5-B**, can be used to determine the input impedance of a transmission line. This was very commonly used before the personal computer became ubiquitous. In addition to the accuracy limitations due to the input and output resolution inherent in a chart, the *Smith chart* assumes that the line is lossless. This may result in significant error depending on the amount of line loss.

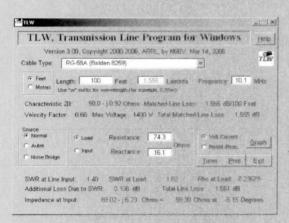

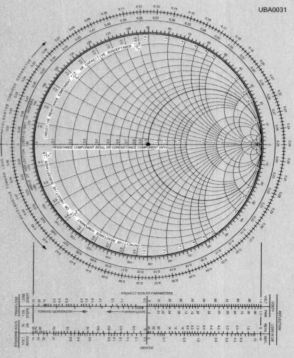

Fig 5-A — The main screen of *TLW* (*Transmission Line for Windows*) software is shown performing an analysis of the L/D of 10,000:1 case from Chapter 4's Table 4-1 at 10.1 MHz.

Fig 5-B — A *Smith chart* used for the graphical determination of the input impedance of a lossless transmission line.

Review Questions

5-1. Describe three reasons it might be desirable to have a transmitter and an antenna in different locations.

5-2. If you want to make a λ/4 section of RG213 for a 10 MHz system, how long would you make it?

5-3. A 1000 W transmitter at 15 MHz is feeding a matched load through 200 feet of RG-8 transmission line. How much power reaches the antenna? Repeat if the frequency is 150 MHz. Repeat both cases if the antenna has an SWR of 3:1.

Chapter 6

Making Real Dipole Antennas

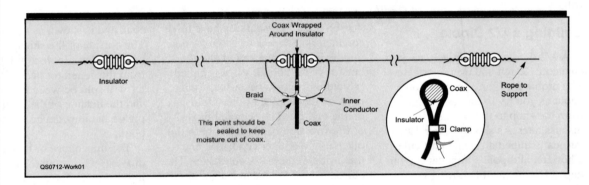

A dipole is easy to make and works well for many applications.

Contents

Making Real Dipole Antennas

Real dipole antennas are not hard to build. If properly constructed and installed they provide excellent performance, especially considering the small investment in materials. For HF use, the most common dipole construction method is a simple wire antenna, as shown in **Fig 6-1**. As discussed in Chapter 4, the actual λ/2 resonant frequency will depend on the thickness of the antenna. For wire this is about 97% of the free-space length. For those working in English units, dividing 468 by the frequency in MHz will provide a length in feet that will be quite close to the needed length.

Building a λ/2 Dipole

A λ/2 dipole is deceptively easy to envision, design and even fabricate. The problems, if any, usually come about as you try to move a design from the shop to the sky, as I'll discuss later. As shown in Fig 4-6, a vertical's impedance, and particularly a horizontal dipole's impedance, will depend on its height above ground. The impedance will depend to a smaller extent on ground conditions. The impedance at resonance will generally range from 40 to 100 Ω, nice values compared to available coaxial cable. You would think that you could just connect the center conductor and the shield to the arms of the dipole and you would have the electrical part done. Not quite!

What About Balance vs Unbalance?

A dipole is inherently a *balanced system*; that is, each arm is equally above "ground" potential. Coaxial cable is inherently unbalanced; that is, the outside of the shield is intended to be at ground potential. What does "ground" mean if we have a dipole, say 50 feet in the air? You could spend a lot of time discussing and pontificating on this topic and many have!

The important thing to understand about coaxial cable is that at RF frequencies, due to the skin effect, you essentially have three conductors. There is a center conductor, the inside of the shield and the outside of the shield. What is called "shielding" occurs because the currents on the outside of the shield stay on the outside. The "coaxial mode" of transmission happens between the outside of the inner conductor and the inside of the outer conductor. The currents there are balanced and that transmission mode is often referred to as *differential mode*. Any unbalanced currents flow on the outside of the coax and such current is usually called *common mode*.

Fig 6-2 is a simple model of the electrical configuration of a dipole fed directly with coax cable. The desired differential mode current is applied to the center through the feed

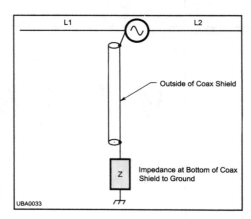

Fig 6-2 — Model of direct connection of coax cable to a dipole.

coax and is shown as a generator. The outside of the shield is shown connected to dipole arm L1. Current from the generator headed towards L1 will split between the antenna arm and the outside of the shield depending on the impedance of each at that point.

The impedance of half a resonant dipole (L1) is around 36 Ω, depending on height above ground. The impedance of the outside of the shield — and thus the amount of current flowing on the shield — will depend on its connection (usually to a ground point) and the length of the shield. In effect this setup looks like another antenna element. Thus, if the impedance to ground is zero and the length of the shield is λ/4, the impedance of the shield will approach infinity.

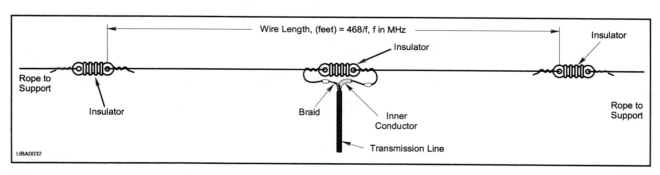

Fig 6-1 — Details of a simple λ/2 wire dipole for HF use.

Little current will thus flow on the shield. Make the coax length λ/2, however, and the impedance at the antenna is closer to zero and more current will head down the outside of the coax.

Even if the above situation is corrected, as discussed below, current can get on the outside of a shield by coupling to the antenna itself. In the ideal antenna installation, the transmission line is perpendicular to the antenna. In that configuration, any current coupling from one side of the antenna to the transmission line is cancelled by radiation from the other side. This effect is less pronounced as the line gets further from the antenna, and many suggest it is not a problem if the line is perpendicular for at least λ/4, but that is not a guarantee that no problems will occur.

What Does Coax Shield Current Do to Us?

This is another topic of frequent discussion. One thing that happens is that any current that flows on the outside of the coax will radiate, just like any antenna. This will reduce the power available for the primary antenna. It will likely have a different polarization and will radiate in different directions than the radiation from the antenna itself. This can be good or bad. If you would like coverage from your dipole in all directions, the chances are that the radiation from the shield will fill in the deep nulls of the dipole off the ends. On the other hand, if you want to concentrate in a particular direction and eliminate transmission and reception from particular directions, the shield current will likely significantly distort your pattern.

A more subtle problem may result from the location of the radiation. Typically, antennas are installed away from systems that might be troubled by strong RF fields. If the coax running through the building is radiating due to shield current, you may get signals to (and from) devices that you don't want to hear. For a dipole over real ground, as for the 10 MHz dipole shown in **Fig 6-3**, the difference with and without the stray radiation appears small, although the nulls are

less deep for the case without a shield. For very directional antennas, the difference can be significant.

Another common problem with current on the shield happens when it reaches the transmitter. The equipment grounding is rarely perfect and as a consequence, the current doesn't go quietly to ground, but rather shows up as an RF voltage on the equipment chassis. This can provide a tingle when your lip touches the microphone, or worse, it gets into transmitter circuitry resulting in feedback or control system lockup.

How do You Keep the Current Where You Want it?

If you insert a high enough impedance in series with the common mode impedance of the transmission line at the antenna, as shown in **Fig 6-4**, you can force the current to flow on the antenna rather than down the coax shield. This can be accomplished with a wideband transformer called a current balun, also known as a choke *balun*. A balun converts a *bal*anced load to *un*balanced coax, or vice versa.

The choke can be as simple as making a coil of the same coaxial cable used to feed the antenna, or it can be as complicated as a specially wound ferrite-core transformer. The latter can provide good transformation over a wider frequency range. A coaxial current choke near the transmitter end can be used to keep any currents coupled to the outside of the coax from getting into the station equipment.

Nuts and Bolts

As shown in Fig 6-1, there are not many electrical complexities associated with the construction of a horizontal HF dipole. The major challenges are mechanical — how do you get the dipole high enough to have the radiation take off at the desired elevation angle? How do you get your dipole to stay there in the presence of wind, weather and flying creatures?

The optimum construction method for an HF dipole is to suspend the center and both ends from fixed sup-

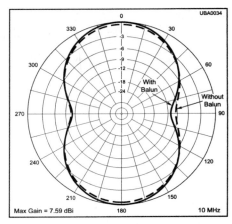

Fig 6-3 — Azimuth pattern of 10 MHz dipoles with and without baluns.

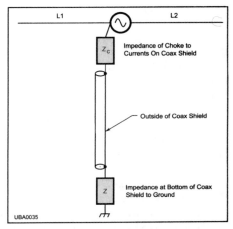

Fig 6-4 — Use of a balun or choke inductance to limit currents on the outside of a coax cable.

ports that do not move in the wind. This results in an unsupported span length of about λ/4 (for a λ/2 dipole). Intermediate supports can also be provided if circumstances warrant. Note that the span will sag somewhat, unless an infinite amount of tension could be applied to the halyards suspending the span. This clearly is not possible and thus you will end up with a reasonable amount of sag. Standard civil engineering techniques can be used to determine the amount of sag for a given length, wire weight and tension.

You can also design the dipole to make sure that the recommended tension for the size wire is not exceeded to determine the resulting sag. **Table 6-1** specifies the maximum suggested tension for two commonly used antenna wire types. The result-

Table 6-1

Maximum Tension for Stressed, Unloaded, Antenna Wire.

American	Recommended Tension[1] (pounds)		Weight (pounds per 1000 feet)	
Wire Gauge	Copper-Clad Steel[2]	Hard-Drawn Copper	Copper-Clad Steel[2]	Hard-Drawn Copper
4	495	214	115.8	126.0
6	310	130	72.9	79.5
8	195	84	45.5	50.0
10	120	52	28.8	31.4
12	75	32	18.1	19.8
14	50	20	11.4	12.4
16	31	13	7.1	7.8
18	19	8	4.5	4.9
20	12	5	2.8	3.1

[1]Approximately one-tenth the breaking load. Might be increased by 50% if the end supports are firm and there is no danger of icing.
[2]Copperweld,™ 40% copper.

ing sag in the center of the half span (see **Fig 6-5**) for a given amount of tension can be estimated using the nomograph shown in **Fig 6-6**.

Many other mechanical configurations have been used, with the most common perhaps being the dipole slung between two trees. This is often constructed by using some kind of line launcher, such as a sling shot, fly-casting rod, bow and arrow or even a radio-controlled helicopter to get the line over a desired tree limb. While this method has been used successfully many times, almost all such installations eventually fail due to a combination of factors. The major problem is that the trees move in the wind. If the trees at both ends moved the same distance, in the same direction and at the same time, the tension would remain constant. Unfortunately, trees can't be counted on to do any of those things.

Keeping it Up in that Tree

To maximize the success of a tree-mounted dipole, you must take tree sway into account. One way is to note the maximum sway of the trees at each end and add that length to each halyard once set for the appropriate sag to give the maximum tension specified in Table 6-1. Note that if there is no center support, and the weight of the transmission line is thus also supported by the antenna, additional tension will be applied to the wire and you must allow for ad-

ditional sag.

An additional failure mode is halyard *chafing*. As the tree moves back and forth, the bark cuts into the halyard rope, eventually causing it to break. Chafing can be minimized if the halyard is run directly down and secured to the trunk of the same tree that it's over. In that way the halyard tends to move with the tree rather than having the tree move along the halyard.

Some rope is better than others at resisting chafing. Twisted nylon rope has excellent stretch characteristics; however, it seems to chafe very quickly. I have found "yacht braid," a generic name for rope used for sailboat halyards, a rope that lasts quite long in antenna-mounting service. Although it is designed for low stretch, it makes up for that deficiency by resisting chafing.

I suppose the strongest halyard material is stranded aircraft-grade stainless-steel rope, although it may saw through the supporting tree limb in place of chafing. One trick to reduce chafe breakage is to start with an extra long halyard. Every year or so, lower the antenna end and inspect for chafing. If it has started, cut off a few feet of rope at the antenna end and re-hoist. You will move the chafed area away from the limb.

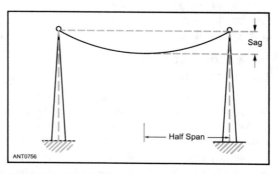

Fig 6-5 — The half span and associated sag of an unloaded wire suspended from both ends.

Some have used stretching elements at the antenna insulators. Some insulators are made with internal springs; however, their travel range is usually insufficient to solve a serious tree-sway problem. Some have employed screen door closing springs for the purpose. Another approach is to have a sacrificial halyard section of lighter line holding the ends of a loop of the main halyard. If tree sway strain happens, the light line will break and the loop inserts enough extra halyard to avoid excessive tension. A combination of these methods can also be used.

A better way is to use the pulley arrangement shown in **Fig 6-7**. This solves all the problems at once. There is no chafing because the rope holding up the pulley is fixed (as long as it is secured to the same trunk, as noted previously). A constant tension is applied by the weight, independent of tree sway. You can adjust the tension easily by adjusting the weight used.

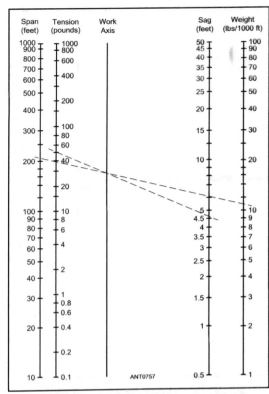

Span (feet)	Tension (pounds)	Work Axis	Sag (feet)	Weight (lbs/1000 ft)

Fig 6-6 — **Nomograph for determining wire sag. (John Elengo, Jr, K1AFR)**

For all halyard arrangements, you should provide an extra *downhaul* line from the halyard side of the end insulator. This line is used to get the end of the halyard down in the event the antenna breaks. There's nothing more frustrating than completing a five-minute antenna repair and then spending the whole day getting the halyard back up in the tree!

A Special Case — the Inverted V Dipole

Finding two or, even better yet, three solid supports for a dipole is often difficult. Fortunately, an antenna with similar, but not quite equal performance, can be constructed with a single support by putting the apex of the antenna on top of the support and extending the dipole arms at an angle downward to be secured near the ground. **Fig 6-8** shows the configuration. In addition to needing only a single support, this arrangement has the advantage that the support can also support the weight of the center structure and transmission line.

The downside to the *inverted V dipole* is that the maximum radiation occurs at a somewhat higher angle than for a horizontal dipole at the same height. See **Fig 6-9**. It shows the elevation pattern of a 10 MHz λ/2 flattop dipole at 50 feet over typical ground, compared to the pattern for an inverted V with an included angle between the two legs of 90°, with its apex at 50 feet, the same height as the flattop dipole. The V has its maximum radiation at 31° rather than 27° for the dipole. The 50 foot dipole could be lowered about 12% to 44 feet and have the same elevation response. The inverted V also has about 1 dB less gain at its maximum elevation. In many applications, these are reasonable trade-offs.

For the same wire length, a 90° inverted V will have a resonant frequency about 6% higher than determined by the 468/F formula, depending on the droop angle from the horizontal. Shallower angles, say 120°, will be closer to the performance of a horizontal dipole, and steeper drooping angles could also be used, with correspondingly reduced performance. One advantage of the Inverted V configuration is that the real part of the feed-point impedance drops somewhat as the dipole arms are lowered and as a result the match to commonly used 50 Ω coaxial cable is improved.

Vertical Dipoles for HF

We have been concentrating on horizontal HF dipoles so far because they are the most frequently encountered HF dipoles. Vertical HF

antennas are more often of the *monopole* variety (think half a dipole), because they can be easily fed at the base. We will discuss monopoles in another chapter later in the book.

Vertical half-wave HF dipoles are, however, sometimes encountered. The primary advantages of a vertical dipole over a horizontal one are that they provide omnidirectional azimuth coverage and, like an inverted V, they require a single support — *almost*. In order to work most effectively, the feed line should be perpendicular to the antenna for at least λ/4. This generally requires a second support at least half as high as the primary one as shown in **Fig 6-10**. An additional advantage is that they provide low angle coverage, even at low heights. The major support must be at least λ/2 high for a full-sized λ/2 vertical dipole, and the actual low angle response of an inverted V on the same support may have a similar pattern.

A potential disadvantage of a typical vertically polarized antenna compared to a horizontal one is that the reflected signal from a horizontal antenna above real ground tends to add significantly to the signal at medium elevation angles. Over perfect ground, the reflected in-phase signal from a vertical theoretically should add significant power at very low takeoff angles; however, this reflected power is lost for waves traveling over typical lossy soil. Now, if you have

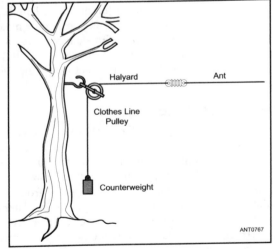

Fig 6-7 — **Pulley arrangement for securing an antenna halyard to a tree while avoiding chaffing and high tension during tree movement.**

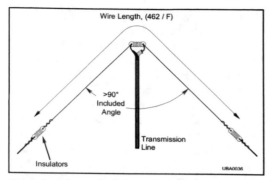

Fig 6-8 — Inverted V wire dipole.

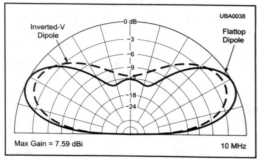

Fig 6-9 — Elevation pattern of inverted V wire dipole (dashed) compared to that of horizontal dipole (solid)

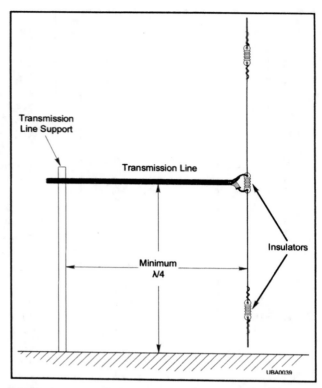

Fig 6-10 — Vertical HF wire dipole showing need for second support to maintain perpendicular transmission line.

seawater in the desired directions, a vertical would have much more reflected signal at low elevations than even a higher horizontal dipole. But for most types of lossy soil, the situation is as shown in **Fig 6-11**.

What About Other Lengths?

A dipole with a length of approximately $\lambda/2$ is often used because it provides a convenient impedance match to popular transmission line types. As noted in Table 4-1 and Figs 4-3 to 4-5, dipoles of other lengths can also be used, but the feed-point impedance will be quite different. These lengths can also be used effectively, as long as an appropriate matching network is provided to match the impedance to that required by the radio equipment or the transmission line. I will discuss some particular applications of different length dipoles in Chapter 12.

VHF and UHF Dipoles

I have been focusing on HF dipoles so far. However, a dipole also works well at higher frequencies. As a matter of construction convenience, dipoles above the HF range tend to be made from solid tubing rather than flexible wire. Rigid tubing has the advantage that it can be supported from the center insulator and thus doesn't require support or insulators at the ends. In many cases, since their size makes it so easy, at VHF and UHF dipoles are parts of directional arrays, a topic for later discussion.

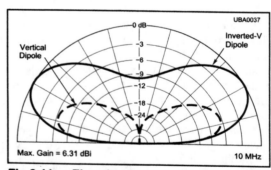

Fig 6-11 — Elevation pattern of vertical 10 MHz dipole suspended from 50 foot support (dashed) compared to that from an inverted V from the same support in its preferred direction (solid).

Review Questions

6-1 What are two reasons why it is desirable to have a support at the center of a wire dipole?

6-2 What are the results of changing from a thin coaxial cable to a thicker one with lower loss? Case 1 with no support at the center of the dipole; case 2 with a support?

6-3 Why is it better to secure an antenna halyard at the base of the same tree that the halyard is over?

Chapter 7

The Field From Two Horizontal Dipoles

Chuck Hutchinson, K8CH, adjusts his two element array.

Contents

The Field From Two Horizontal Dipoles

In Chapter 3 I described the action of a dipole in the presence of its below-ground "image antenna." It is also common to find antenna arrays of multiple real dipoles. As you might expect from the discussion of the dipole above its image antenna in Chapter 3, horizontally polarized antennas combined above one another will result in a change of the elevation pattern, while side-by-side antennas will change the azimuth pattern. This chapter will begin to examine how two dipoles can be combined to achieve some desired outcomes.

Horizontal Dipoles in a "Stack"

Perhaps the easiest configuration to start with is to add an additional dipole above the one you used in the last few chapters and find how it performs. A broadside view of the configuration is shown in **Fig 7-1**, with an end view shown in **Fig 7-2**. It looks very much as you would expect, but note that unlike the image configuration in Chapter 3, I chose here to feed both antennas in the same phase. Unlike the image situation, this is a choice depending on how you connect the sources. I could have connected each antenna to sources that are out-of-phase.

Another difference of note is that the height of this two-element array (and other large antenna structures) is taken as the height of the center of the array. That was true of the single dipole as well, but for the single dipole the height also was as high as the antenna got. This may have ramifications on the size of support structures and clearances, so keep it in mind.

If you divide the available transmitter power in half and apply each half to the two dipoles in free space, each will separately attempt to radiate the same kind of signal I described in Chapter 2.[1] However, since the signals are on the same frequency, when you look at the response you see the combination of the output from both antennas, not just one.

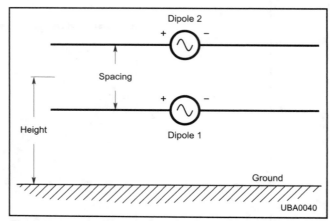

Fig 7-1 — Front view of configuration of two in-phase stacked horizontal dipoles.

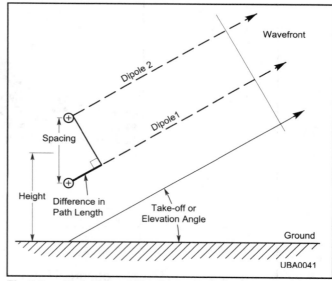

Fig 7-2 — End view of the two horizontal dipoles with polarity as indicated.

So What Happens?

The first antenna configuration that I will examine is one in which the dipoles are spaced λ/2 apart. Because the dipoles are driven in the same phase, every point in space that is the same distance from each antenna will receive the benefit of each antenna's pattern adding together. This clearly happens going horizontally away from the plane of the antennas (for any spacing), as shown in **Fig 7-3**. This is the direction broadside to the plane of the array (imagine the antennas stapled to a billboard) and thus this configuration of dipoles is often called a *broadside array*. Note that if you feed them in the opposite phase the signals in the broadside direction would cancel, just as in the case of a horizontal antenna and its image — more about this later.

A single dipole in free space will radiate the same level signal at all elevation angles. As seen in Fig 7-3, this is not happening with two dipoles. Let's consider the radiation from one, say the lower, dipole that would head upwards if left to itself. The radiation going upwards, towards the second dipole, starts out in the same phase as is fed to the second dipole, but by the time the signal has traveled the λ/2 towards the upper dipole, the signal in the upper dipole is exactly 180° out-of-phase. Since the signals in that direction are the same strength, they combine by cancelling out going upwards! You have made an antenna with a *null* in the pattern at a 90° elevation angle. Note that for the free-space case, the exact same thing happens going downward.

What Happens to All the Energy?

Note that nowhere did I talk about anything getting warm or otherwise dissipating energy. All the energy fed to the antenna system will be radiated. If it doesn't go one place, it will go another. Just as the light from a flashlight bulb with a reflector is brighter going out the front, and weaker elsewhere, the energy from this broadside antenna is greatest perpendicular to the "billboard" (unlike the flashlight, on both sides of

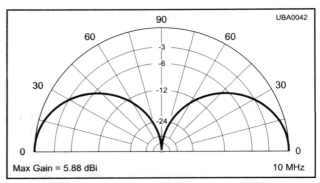

Fig 7-3 — *EZNEC* elevation plot for two element horizontal array in free space.

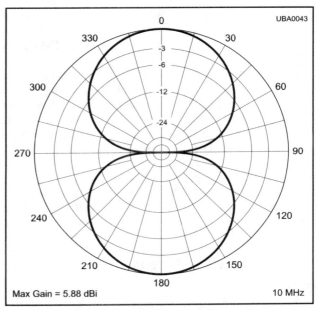

Fig 7-4 — *EZNEC* plot of azimuth pattern of two element array in free space.

the billboard) and less elsewhere. If you compare the field strengths at the maximum of the elevation patterns of the two-dipole array (Fig 7-3) with that of the single dipole in free space (Fig 2-8) you will find that the difference in maximum signal is 5.88 − 2.06 = 3.82 dB.[2] This represents an increase in power by a factor of 2.41, a bit more than double that from the single dipole in the maximum direction.

Someone listening from the direction of maximum radiation (the main *lobe*), if they didn't know about the exact configuration of the antenna, could think you more than doubled the transmitter power. This effect

illustrates the concept of *antenna gain*, although there really isn't any *more* power, you have just moved it around, concentrating the power more in certain directions.

Note that someone from right above the antenna might think you turned off your transmitter! Note also that the azimuth pattern (see **Fig 7-4**) remains largely unchanged.

What If We Get Back Down to Earth?

While the free-space pattern is a good way to isolate the effect of adding the second antenna, the real-world situation is near the ground. As you might expect, you still have the effect

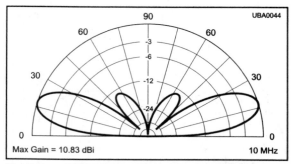

Fig 7-5 — *EZNEC* plot of elevation pattern of two element horizontal array over perfect ground.

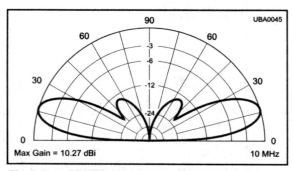

Fig 7-6 — *EZNEC* plot of elevation pattern of two element horizontal array over typical ground.

of the reflections from the ground to deal with, even though you don't have any signal heading straight downward from the two stacked dipoles in Fig 7-2.

No matter how complex the array, the effect of the ground is still that of an image antenna, 180° out-of-phase with the real horizontally polarized antenna, and the same distance below the ground. The result for a horizontally polarized antenna is still a cancellation of energy at 0° elevation and a reduction of radiation at very low angles, as shown in **Fig 7-5** for perfect ground and over typical ground in **Fig 7-6**.

Let's Feed the Two Antennas Out-of-Phase

So far you've been looking at the case where the generators attached to the antennas are of the same phase. This is a useful and frequently encountered case, but clearly only one of many ways you could drive the antennas. The other extreme case is to turn one of the generators around so that the two antennas are fed out-of-phase. If you think about it, this is similar to the horizontal antenna and its image, but in this case both elements are above ground.

For the case of λ/2 spacing, the wave from the lower antenna starts out at the opposite phase from the signal in the upper antenna. By the time it propagates to the upper antenna, the phase of the signal in that antenna has shifted exactly 180° and now the two signals propagate upward in the same phase and thus they add, as seen by any receiver directly above. The

signal heading downward has the same result and another lobe, at least in free space, heads downward. The signals perpendicular to the array, the broadside direction that had the maximum response previously, now has a null. The name often given to this configuration is somewhat confusing, since it is called an *end-fire array*. A more descriptive name might be *edge-fire*, or something else that doesn't make it sound like the signal is leaving the ends of the elements. After all, the azimuth pattern still has a null off the ends of the dipoles. **Fig 7-7** is the elevation pattern in free space for the 180° out-of-phase case.

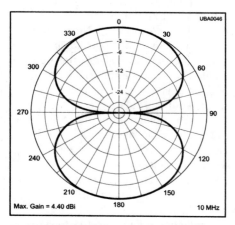

Fig 7-7 — *EZNEC* elevation plot of two element end-fire array of vertically stacked elements in free space.

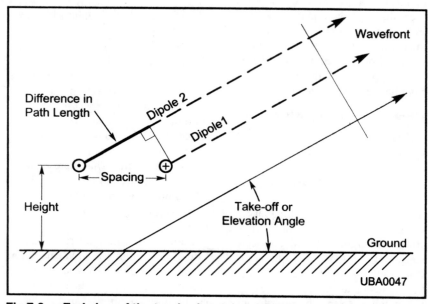

Fig 7-8 — End view of the two horizontal end-fire dipoles, with phases as indicated.

The End-Fire Array on Its Side

An interesting case to consider is the end-fire array with both antennas at the same height, and still fed 180° out-of-phase as shown in **Fig 7-8**. An *EZNEC* elevation plot over typical real ground is shown in **Fig 7-9**, and it is interesting to compare this with the similar looking plot of the broadside array response in Fig 7-5. This is an indication of an important general principle — there are often multiple configurations that will provide the response you want. You can pick the one that best suits the physical limitations of a particular installation.

What if We Use a Different Dipole Spacing?

The λ/2 spacing cases just examined represent a very special distance in which the radiation in the plane containing the elements cancels or adds because the time it takes a wave to move between the two elements results in exactly a 180° phase difference. There is nothing to prevent you from picking any other spacing, and there may be reasons why other spacings are beneficial. One extreme would be a spacing of zero, or one dipole resting on the other. If the two are fed in the same phase, they just act like a single, but slightly thicker antenna. If you attempt to feed the two dipoles out-of-phase, it will act as if you short-circuited the transmitter and you will not be able to put any energy into the system.

At other spacings you will find that the shape of the response gradually changes. At smaller spacings the cancellation in the previous null directions is less than 100%; while at larger spacings there can be additional peaks and nulls because there are path differences greater than λ/2. I will examine the effect of spacing in more detail in the chapter on Arrays of Multiple Elements.

A Unidirectional End-Fire Array

A very interesting special case of the end-fire configuration exists if you change the spacing to λ/4 and the phase between the elements from 180° to 90°. With this arrangement, as shown in **Fig 7-10**, with a 90° delay (we'll show how to do that soon) to the signal connected to Dipole 1, its signal has advanced to 0° by the time the signal from Dipole 2 gets there and they add in-phase in the direction from Dipole 1 towards Dipole 2. On the other hand, the signal from Dipole 2 at –90° reaches Dipole 1 when it is at +90°, cancelling in that direction. See **Fig 7-11** for the elevation-plane plot of this array and **Fig 7-12** for the azimuthal pattern.

It should be clear that there are many applications in which a unidirectional antenna would be useful — for example a radar operator could tell if an aircraft is in front of or behind the radar. As I

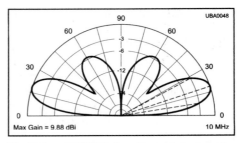

Fig 7-9 — *EZNEC* elevation plot of end-fire array 0.75λ above real ground.

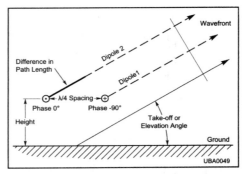

Fig 7-10 — **Physical configuration of two element unidirectional array.**

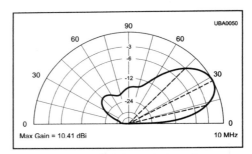

Fig 7-11 — *EZNEC* plot of elevation pattern of two element unidirectional array over typical ground.

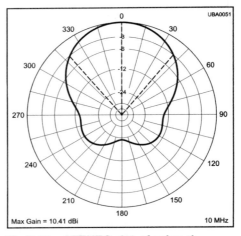

Fig 7-12 — *EZNEC* plot of azimuth pattern of two element unidirectional array at peak of elevation pattern.

will discuss later, this is just one of many ways in which you can generate a unidirectional pattern.

Two Dipoles in Line — Collinear arrays

Another configuration of two dipoles is to have them along the same line, end-to-end. The configuration is shown in **Fig 7-13**, with the path difference in the signals shown in **Fig 7-14**. Note that the spacing can be any length larger than λ/2, since they would overlap if less, and then they would not act as two separate dipoles. For the case of the minimal λ/2 spacing at a height of λ/2 over real ground, the azimuth pattern is shown in **Fig 7-15**. In the direction broadside to the two dipoles, the patterns add, as you would expect, but at any other azimuth the propagation distance is different, as shown in Fig 7-14.

The further you move away from broadside, the greater the difference in the path length, with the result that the patterns from the two antennas no longer reinforce each other as you shift off axis. Unlike the stacked case, you have thus modified the azimuth pattern, but not the elevation pattern (compare **Fig 7-16**, the elevation pattern, with Fig 3-7).

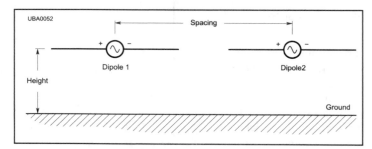

Fig 7-13 — Front view of two element collinear array.

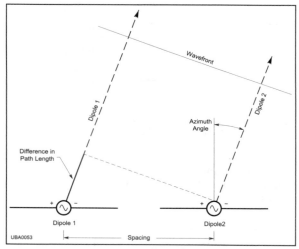

Fig 7-14 — Physical configuration of two element collinear array.

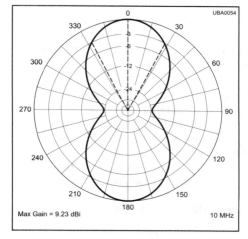

Fig 7-15 — EZNEC plot of azimuth response of two element collinear array.

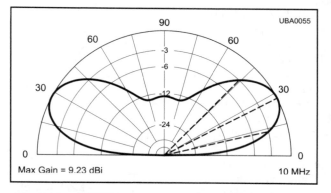

Fig 7-16 — EZNEC plot of elevation pattern of the collinear array over real ground.

Chapter Summary

In this chapter I have introduced and discussed the topic of how two horizontally polarized dipoles can be used to deliver more RF energy to smaller volumes of space than a single dipole would. If you stack the dipoles one above the other, you limit the elevation coverage. In a special case you can limit the coverage to a single direction.

If you put the dipoles side-by-side, you do not change the elevation coverage, but limit the azimuth coverage. Either arrangement can be used to advantage, since it makes sense to send the energy to places where it can be used.

Notes

[1]To implement "half-power" in the *EZNEC* model description, each source voltage or current should be set at 0.707 times the full value (usually 1.0) because the power is proportional to the square of the voltage or current.

[2]If you are not comfortable working with decibels, spend some time with Appendix A.

Review Questions

7.1 Under what circumstances might a 2-element broadside array be beneficial?

7.2 Given that broadside and end-fire arrays can have similar responses, why might one be preferable to the other?

7.3 What are some of the benefits of a collinear array, and some of the disadvantages in comparison to the other two antennas?

7.4 Describe two potentially beneficial uses for a unidirectional array.

Chapter 8

The Field From Two Vertical Dipoles

The combined field from multiple antennas can be oriented in many helpful ways.

Contents

The Field From Two Vertical Dipoles

In Chapter 7 I described the way the fields from two horizontal dipoles combine at a distant point. I pointed out that if the two dipoles were fed equal amounts of energy in the same phase that the energy would add everywhere the distance to each of the dipoles was the same. I also noted that the presence of the ground reflection could modify the result, depending on how high above the ground the antenna elements were located. If the elements were fed in opposite phase (180°) the energy would cancel in directions that were equal distances from the elements and thus the directional characteristics would be changed.

As you might expect, similar things happen to dipoles that are oriented vertically. In fact, you may wonder why vertical dipoles rate a whole chapter, since they are essentially the same as horizontal antennas placed on their side. The big difference is in the way that ground reflections act on vertically polarized signals. I will start with essentially the same configurations as in the horizontal case in Chapter 7.

Side-by-Side Vertical Dipoles

The side-by-side case is in many respects similar to the "stack" of horizontal dipoles I started with in the last chapter. A broadside view of the configuration is shown in **Fig 8-1** with a bird's-eye view in **Fig 8-2**. Here both dipoles are fed in the same phase. This means that their fields will add in the direction perpendicular to the plane of the antennas.

By convention, the height of this array is taken as the height of the center of the array. With a vertical λ/2 dipole, this distance cannot be less than λ/4 because otherwise the bottom would be on the ground. I have chosen a height of 36 feet for a 30 meter dipole. This places the bottom 12 feet above the ground. This is a special height, as I will discuss later, but I could have selected any height greater than about 24 feet to keep the bottom of the dipole off the ground.

In-Phase Performance

With equal power to the two elements, the free-space azimuth pattern shown in **Fig 8-3** will result. As expected, the maximum field strength occurs in the direction broadside to the plane of the elements (0° per the indicated azimuth angle in Fig 8-2). You won't be too surprised at the free-space elevation pattern, taken at the maximum of the azimuth pattern — in the broadside direction and shown in **Fig 8-4**.

If you modify the *EZNEC* model by putting the antenna above perfect ground, you will see in **Fig 8-5** the first major change from the horizontal case. Unlike the horizontal array, the vertical array provides the strongest signal at the horizon (0° elevation). Again, this is a result

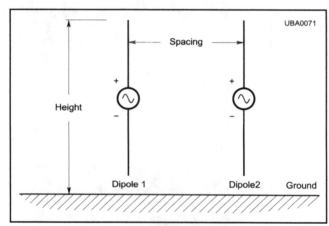

Fig 8-1 — Front view of configuration of two in-phase stacked vertical dipoles.

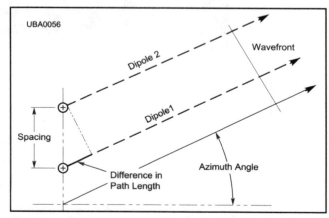

Fig 8-2 — End view of the two vertical dipoles, with polarity as indicated.

of the antenna image (the apparent starting point of the ground reflection) being in-phase with the antenna itself for vertical polarization. Arrays of this sort can approximate the perfect-ground case if there is highly conductive ground in the desired broadside direction. If the antenna array is located adjacent to a body of saltwater, or if many wavelengths of wire mesh are used to simulate a highly conductive ground, very low takeoff angles can be achieved.

Over typical ground, the picture is not quite so rosy. The electromagnetic waves traveling adjacent to the ground heat up the lossy ground resulting in less energy traveling along the surface. A number of other effects combine to eliminate the very low-angle propagation, as shown in **Fig 8-6**, where the patterns for perfect and typical grounds are compared. Compare these results to those of the horizontal array in Chapter 4 at a comparable height and note that the peak of the vertical beam is still at a slightly lower angle.

Different applications often find one orientation superior to another. For example, early naval shipboard radar systems tried vertical polarization, with the hope that the low-angle coverage would be excellent for observing distant surface vessels. What they found instead was that reflections from seawater wave fronts (called *sea clutter* in the radar business) overloaded the receivers and masked real targets. A change to horizontal polarization with its null at the horizon largely eliminated the problem. The height of the usual naval shipboard radar was sufficient that distant targets could still be seen above the horizon.

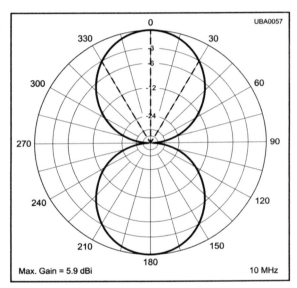

Fig 8-3 — *EZNEC* plot of azimuth pattern of two element array in free space.

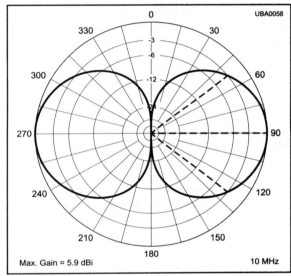

Fig 8-4 — *EZNEC* elevation plot two element vertical array in free space.

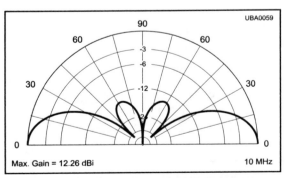

Fig 8-5 — *EZNEC* plot of elevation pattern of two element vertical array over perfect ground.

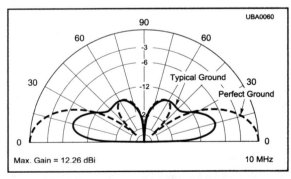

Fig 8-6 — *EZNEC* plot of elevation pattern of two element vertical array over typical and perfect ground.

Vertical Dipoles in a Line

Vertical dipoles in a collinear configuration make a particularly useful antenna system. By placing one dipole above the other, as shown in **Fig 8-7**, and feeding them in phase we reinforce the radiation near the horizon. The resulting elevation pattern is shown in **Fig 8-8** for two collinear dipoles (dashed line) compared to just the lower dipole (solid line). When the second dipole is added, the peak field intensity becomes somewhat stronger and is focused closer to the horizon. This configuration, as well as designs with additional in-phase dipoles, are commonly used by VHF and UHF base stations who need to communicate with vehicle-mounted radios in all directions. At VHF/UHF a number of dipoles can be combined in such a vertical collinear configuration and still be of reasonable size. The additional gain improves the signal-to-noise ratio at the edges of the coverage area.

It is interesting to look at the effect of element spacing on the resultant pattern. The maximum gain of two collinear dipoles occurs at a spacing of 1 λ (λ/2 between element ends). The resulting pattern is shown in **Fig 8-9**. While the peak gain is about 1.7 dB higher than the close-spaced collinear, the energy going upward is starting to become predominant. This is because at wider spacings the path length from the two elements can be in-phase at angles other than the broadside direction.

The peak at about 43° elevation results because there is still enough energy in each element's pattern and they are close enough to being in-phase that they add to a large measure. At 43° elevation, the difference in path length is 0.68 λ (244° compared to 1 λ or 360° for in-phase). At a spacing of 1 λ, the next elevation at which the two waves will be in phase is 90°, or straight-up. But the individual elements don't radiate at 90°. Hence there is nothing to combine and there is thus still a null at the vertical.

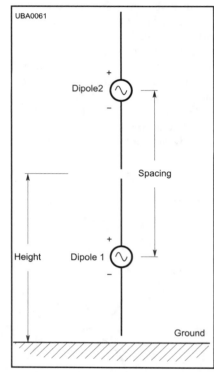

Fig 8-7 — Elevation view of two vertical dipoles in collinear configuration.

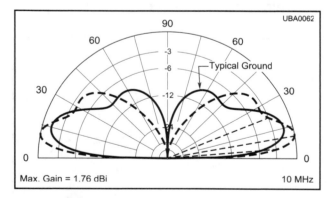

Fig 8-8 — *EZNEC* elevation pattern of two closely spaced collinear vertical dipoles (dashed line) compared to the pattern of just the lower dipole (solid line) above typical ground.

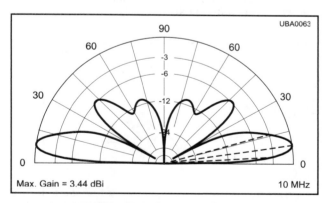

Fig 8-9 — *EZNEC* plot of elevation pattern of two element vertical collinear array with center-to-center spacing of 1 λ.

For any spacings greater than 90° (λ/4) there will be elevation angles at which the path difference is 360° within the pattern of each antenna, so the combined pattern gets more complex and often less useful. For example, **Fig 8-10** shows the pattern at a spacing of 2 λ. At this spacing, the waves from the two dipoles are still in-phase at the horizon, but are in-phase again at an elevation angle of 30°. (See the sidebar for a description of the calculation.) Since the individual dipole patterns are only down a small amount there, a significant additional lobe results.

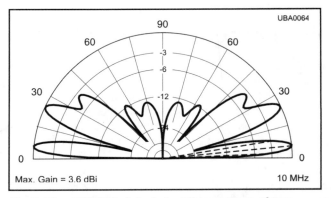

Fig 8-10 — *EZNEC* plot of elevation pattern of two element vertical collinear array with center-to-center spacing of 2 λ.

Phase Steered Arrays

The two element vertical array can be used as an antenna system that can be *steered* to point in different directions by changing phase. While the λ/4 spaced horizontal array can have its direction switched by changing the sequence of feeding the two dipoles, a vertical array offers more possibilities. It should be easy to imagine a pair of λ/2 spaced vertical dipoles providing either east-west or north-south coverage, depending on whether they are fed in-phase or out-of-phase. A possibly even more useful case is the λ/4 spaced case, where either of two unidirectional patterns can be had along the plane of the wires. This is similar to the horizontal unidirectional arrangement of Chapter 4. In addition, the two elements can be fed in-phase to provide broadside bidirectional coverage, at somewhat less gain than the more wide-spaced case.

Perhaps the most extreme example of an application of phase-steering technology is that of the USAF AN/FPS-115 Pave Paws phased array radar shown in **Fig 8-11**. This system provides a "no moving parts" radar system in which each of the two faceplates (one shown, the other oriented to provide nearly 270° total coverage) are composed of 896 distinct antenna elements, each with its own transmitter and receiver front-end. The array is about 72 feet wide with an operating frequency in the 435 MHz range. By combining the signals from each of the elements with carefully controlled phase, a phase-steered beam with an azimuth and elevation width of 2.2° is achieved, capable of detecting targets up to 3000 nautical miles away. There are many advantages of such a system compared to the usual rotating radar antenna — faster tracking, possibility of multiple simultaneous beams and less wear and tear — to name a few.

Fig 8-11 — View of one panel of Pave Paws phased array system.

Calculating Path-Length Difference

It is relatively simple to calculate the difference in path length between the waves from two antenna elements. I will describe the collinear case, as was just discussed in this chapter. However, the same technique can be used for any configuration of elements.

Since you are looking at the combination of signals at a large distance, you can take advantage of the fact that two parallel lines meet at a point an infinite distance away. You can consider the signals leaving each antenna element on parallel paths as shown in **Fig A**. You also consider that the signals start from the center of each element and that the distance between the two individual wavefronts' starting point is the spacing L.

For an elevation angle of 0°, the two paths are the same length and the difference is zero. As the elevation angle, θ, increases, the difference in path length, D, increases up to the maximum, which occurs in the end-fire case for θ = 90°.

To determine the length, as shown in Fig A, the upper interior angle of the right triangle formed between L and D is the same as the elevation angle (because the sum of the angles in a triangle is 180°). Using plane trigonometry, we can see that the difference in path length is just $D = L \times \sin \theta$.

For examples, look at Fig 8-10 for three elevation

Table A

Phase angle at key elevation angles as a function of elevation angle for 2 λ element spacing.

θ	sin θ	D = L × sin θ	Phase
0°	0	0 λ	0°
14.5°	0.25	0.50 λ	180°
30°	0.5	1.0 λ	360°

angles and see what the path length differences are, as well as the resulting differences in phase between the two signals. (Remembering that a full wavelength results in a shift of 360°) L = 2 λ. As shown in **Table A**, the signals are exactly in-phase at 0°, and thus would add at 0° elevation were it not for ground losses. At 14.5° elevation, the signals are 180° out-of-phase and thus cancel, leaving the null shown. At an elevation of 30° the signals are 360° out-of-phase, the same as being in-phase, so they add again. At 30° off axis, each dipole response is starting to fall off, so the maximum signal occurs at a slightly lower angle.

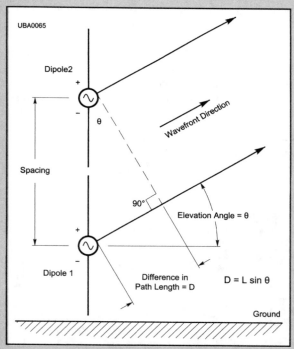

Fig 8-A — Difference in path length of signals from two elements of collinear array.

Review Questions

8.1 Why are arrays of vertical dipoles usually used as base stations for mobile networks?

8.2 Discuss some of the benefits of vertically polarized antenna systems compared to horizontal ones. Consider physical characteristics as well as electrical ones.

8.3 Why might you not want to use vertically polarized antenna systems to communicate with nearby aircraft?

8.4 What do you think would happen as you added more and more vertical collinear elements above the first two?

8.5 Under what circumstances might a conventional mechanically rotating array have advantages over a phase steered system, such as Pave Paws?

Chapter 9

Transmission Lines as Transformers

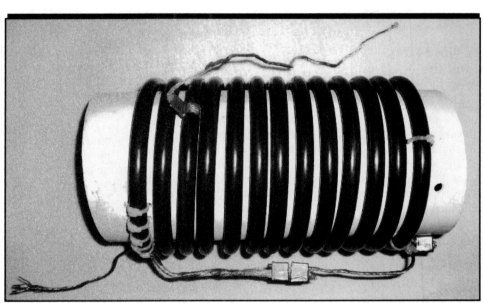

Transmission lines can serve in many roles. Here one is used as a resonant circuit.

Contents

In discussions so far, I have mostly been talking about transmission lines feeding terminations matched to their characteristic impedance (Z_0). It is often the case, either by design or accident, that the load impedance does not match the line Z_0. As I mentioned in Chapter 5, the impedance seen at the input of a transmission line is a function of the load impedance, the line characteristic impedance, the loss in the line and the number (or fraction) of wavelengths of line.

Lines With Unmatched Terminations

It is often beneficial to operate transmission lines into matched loads. The line loss will be lowest for that case, voltages and currents are easily predictable, avoiding stress on components connected to the line. There are cases in which operating lines in an unmatched condition can provide just what we need — easily predictable impedance transformation.

The Quarter-Wave ($\lambda/4$) Transmission-Line Transformer

Along a transmission line that is not matched to its load, the voltage and current will vary with distance, providing a load to the transmitter end that is generally neither that of the far end Z_L, nor the Z_0 of the transmission line. The ratio of maximum voltage on the line to minimum voltage on the line is called the *standing wave ratio* or SWR. A matched line has an SWR of 1:1, a 50 Ω line terminated with a 25 or 100 Ω load will have an SWR of 2:1. There are a whole family of complex impedances that will also have a 2:1 SWR, by the way. However, the computation is much easier with resistive loads.

For a lossless line (and approximately for a short section of real transmission line) the SWR will remain the same along the length of the line. The load impedance, resistive or complex, repeats every $\lambda/2$. The impedance goes to the opposite extreme at odd multiples of $\lambda/4$. For example, a 25 Ω load would get transformed to 100 Ω with $\lambda/4$ or $3\lambda/4$ transmission line sections and vice versa. This effect can be used to your advantage if you wish to transform impedances at

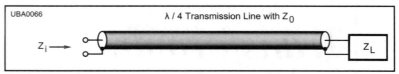

Fig 9-1 — **Quarter-wave transmission-line transformer.**

a specific frequency. For example, if you have an antenna with an impedance of 100 Ω and you connect it to a $\lambda/4$ section of 75 Ω coax, the SWR will be 100/75 or 1.33:1. At the other end of the $\lambda/4$ section, the impedance will be 75/1.33 or 56 Ω, a close match (SWR of 1.12:1) to 50 Ω transmission line. You could then use standard 50 Ω coax cable for a long run to your 50 Ω transmitter with lower losses than if you had used the original 100 Ω load directly, with its 2:1 SWR.

The general expression for the input impedance (Z_i) of a $\lambda/4$ section of transmission line with characteristic impedance Z_0, terminated in a load of Z_L is:

$$Z_i = \frac{Z_0^{\,2}}{Z_L}$$

To solve for the needed transmission line Z_0, we rearrange as:

$$Z_0 = \sqrt{Z_i Z_L}$$

A few observations are in order. While the quantities are shown as complex numbers, transmission lines are available (for practical purposes) only with real, non-reactive Z_0. This implies that the formulation works best for real values of Z_L, as in the above example. This is not a limitation if you work with resonant antennas, which by definition offer a resistive load. Note, as I will

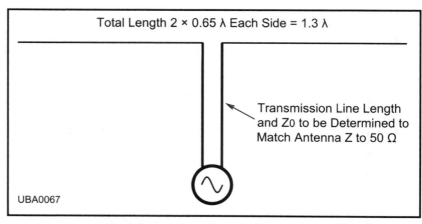

Fig 9-2 — **Extended double Zepp with matching section of line.**

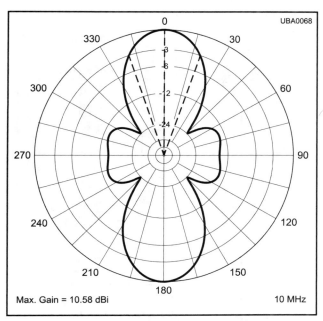

Fig 9-3 — Extended double Zepp azimuth pattern.

discuss, complex impedances can be transformed to match real transmission lines by using transmission line lengths other than λ/4.

The utility of the λ/4 transmission-line transformer method can be limited by the range of available transmission line impedances. Coaxial cable is generally available with Z_0 from about 30 to 90 Ω, while balanced line is available from about 70 to 600 Ω. It is practical to connect transmission lines in parallel (particularly coax) to obtain lower values of Z_0. For VHF and UHF work, it is also feasible to fabricate transmission-line sections with a wider range of values, or non-standard values, through the use of available copper or aluminum tubing.

Since the transformation will repeat with any odd number of λ/4 sections, it is tempting to consider using 3λ/4 or higher odd multiples of a λ/4. This has two potentially significant pitfalls that should be evaluated first. The use of longer mismatched sections will result in higher losses. Often more serious, for sections Nλ/4, the change in Z_i with frequency will happen N times more rapidly with the longer section, effectively reducing the operating bandwidth of the system.

It is also possible to use successive λ/4 sections of line with different Z_0 to work from one impedance to another in steps. This method shares the pitfall described above.

Transformation With Other Lengths of Transmission Line

While a λ/4 transmission line transformer is very useful, especially for resistive loads, sometimes other situations require transformation. A good example is to consider a popular antenna with a reactive input impedance, the extended double Zepp. This antenna is just a center-fed dipole with each side of length 0.625 λ. It is the longest type of simple antenna that can still focus its energy into a single lobe on each side. The extra length and spacing, similar to the two-element collinear, but with a simpler feed system, result in a bi-directional pencil-sharp pattern with higher gain than a dipole. This can be useful to work stations in a particular direction.

While you could perform a lot of math and determine the impedance at the center of this antenna, you can more simply model it using *EZNEC*. The resulting azimuth pattern is shown in **Fig 9-3**, while the impedance plot is shown in **Fig 9-4**. Note

that the impedance at the center at 10.0 MHz is 237.7 – j 1187 Ω, not suggestive of a standard transmission line! Again, avoiding tedious math, you could turn to the *TLW* program and insert the antenna impedance in the LOAD window, using the minus sign to indicate – j. Next, adjust the two available parameters, the line length and Z_0, until you find something that looks like the desired input impedance in the IMPEDANCE AT INPUT indication at the bottom. It takes a few trials to see a direction, but you can find a reasonable length of easily available open-wire 600 Ω line that transforms the impedance to 50 Ω for a coax run to the transmitter. See **Fig 9-5**.

I was lucky to find an easy solution to this matching problem, perhaps a reason that this is a popular antenna. This will not always be the case. If this technique doesn't result in a result that is easy to deal with, sometimes an intermediate value can be found so it can be transformed to your desired impedance through a λ/4 transmission-line transformer. Even better for this antenna might be to use a λ/2 long 4:1 balanced-to-unbalanced transformer as shown in the next section. It is also possible to try various nonstandard, but buildable line sections using the USER DEFINED TRANSMISSION LINE option in *TLW*. See Figure 9-5.

The Half-Wave Loop 4:1 Balanced-to-Unbalanced Transformer

As discussed in Chapter 6, many antennas are naturally balanced with respect to ground, while coaxial

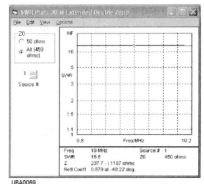

Fig 9-4 — *TLW* impedance and SWR of extended double Zepp.

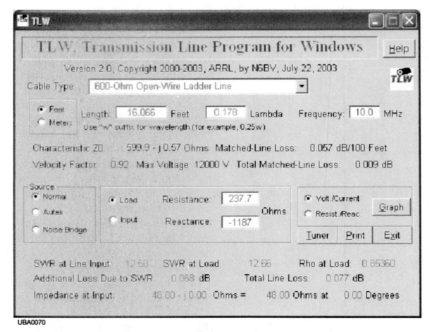

Fig 9-5 — Impedance of extended double Zepp transformed through empirically determined line section.

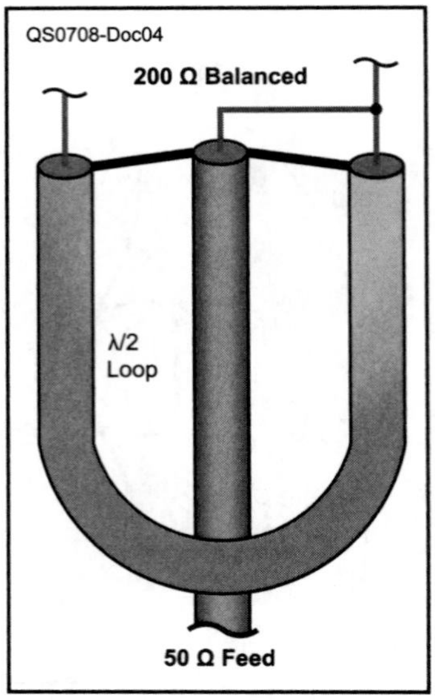

QS0708-Doc04

200 Ω Balanced

λ/2 Loop

50 Ω Feed

Fig 9-6 — Coaxial loop 4:1 transformer-balun.

transmission lines are naturally un-balanced. Sometimes the resulting discontinuity obtained by hooking an unbalanced line to a balanced antenna can cause problems of various types.

In Chapter 6, I mentioned the choke balun as a way to avoid the effects of the unbalance. Another way is through the use of a coaxial cable loop balun, as shown in **Fig 9-6**. The secret to how this works is that a λ/2 of transmission line repeats at its input whatever the load impedance is at the output. It is often possible to arrange an antenna impedance to be a 200 Ω balanced load. As shown in Fig 9-6, half of the load, one side at ground or common potential, appears on one end of the loop. At the other end of the loop it is 100 Ω again and in parallel with the 100 Ω of the other side. Because of the phase difference of 180°, due to the λ/2 length of line, the two sides are out-of-phase and

can drive the antenna properly. The two combine to form a 50 Ω load with one side at ground potential — a perfect spot for connection to 50 Ω coaxial cable.

Note that the transformation ratio of 4:1 will happen for any load and any cable Z_0. The closer the cable Z_0 is to the half load, the less rapidly Z_i will change with changes in frequency, or equivalently, the less fussy the balun will be about cutting accuracy.

Review Questions

9-1 You wish to feed two 50-Ω impedance antennas in phase by joining their transmission lines together. How might you transform the resulting 25-Ω to provide a close match to 50-Ω transmission line?

9-2. Repeat question 9-1 in a different way so that the impedance of the connections of the two antenna feeds is 50 Ω. (Hint — this will require two transformers.)

9-3 Consider a λ/2 transmission-line loop balun constructed from RG-58 coax cable and designed to feed a 10-MHz antenna with an impedance of 200 Ω. What will be the loss in the length of cable considering the resulting SWR?

9-4 Repeat 9-3 for RG-213 cable?

Chapter 10

Practical Two Element Antenna Arrays

This two element array uses loop elements.

Contents

It's not much more difficult to construct two element (or even more) antenna arrays than to construct the dipoles described in Chapter 6. In this chapter, I will discuss a few classic two element antennas, and even toss in a simple four-element array to set the stage for more complex antennas.

The Two element Broadside Array

If you suspend two dipoles, as in Fig 6-1, one above the other, you will make a simple broadside array. As I discussed in Chapter 7, by feeding the two in-phase, you focus the radiation towards the horizon (for free space) or for real ground move it more in that direction.

Fig 10-1 shows the configuration. Note that I have defined the height (h) as that to the array center, and the spacing (s) as the separation between the two elements. You can see the benefits of the configuration with λ/2 spacing by looking at the solid trace in **Fig 10-2** that shows the elevation pattern of the array over typical ground. A dipole also at a height of 50 feet is shown in the dashed trace. Since you need a support of about 75 feet to have the center at 50 feet, it seemed fair to also show a single dipole at that height. Its elevation pattern is shown in the dotted trace. Note that even though the peak of the dotted trace is at a lower elevation angle than the peak of the broadside array, the actual radiation at every low elevation angle is stronger with the broadside array. The array has nulled out the vertical radiation, a benefit in some cases, especially if noise tends to appear from high elevation angles, as it often does.

If it were free space, without ground reflections that cancel at the horizon, the main lobe of the array would be more than 3 dB stronger than either dipole. In this real-world case the improvement is a bit less than 2 dB, still worth doing.

Putting the Broadside Array Together

Putting together a broadside array is something like putting up two dipoles. The main trick is to make sure

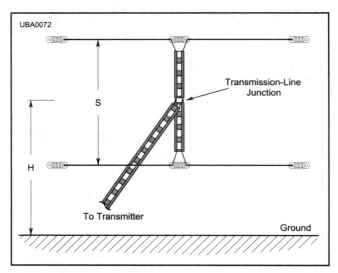

Fig 10-1 — Details of a simple two element HF broadside array.

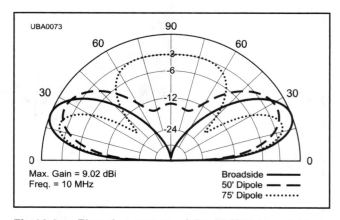

Fig 10-2 — Elevation pattern of the 10 MHz broadside array (solid) compared with dipoles at 50 feet (dashed) and 74.6 feet (dotted).

that you are actually feeding the pair in-phase. This can be accomplished if:

- The transmission lines from the junction in Fig 10-1 to each dipole are the same type and the same length

- The connections are made in the same direction. That is, both right-hand dipole arms must be connected through the lines to the same point at the junction. If they are reversed the dipoles will be 180° out-of-phase — a possible configuration, but not what you want here.

Hooking Them Up

An interesting property of two parallel antennas is that their coupling due to proximity results in a mutual impedance that combines with each antenna's self-impedance to result in a new impedance at each feed point. This effect will be found to some extent for all multielement arrays. For a broadside array, the effect is that each dipole's impedance (that by itself would be about 65 Ω at this height above ground) will be raised to around 80 Ω by the effect of mutual coupling. If you were to feed each antenna with λ/4 of 50 Ω coax, the impedance transformation due to the λ/4 line discussed in Chapter 9 would result in an impedance at the bottom of each line of 31.25 Ω. This results in 15.6 Ω when they are connected in parallel. The result is a fairly high (3.2:1) standing wave ratio for the run to the transmitter if 50 Ω coax is used.

In Search of a Better Match

There are a number of ways to connect the antennas and achieve a reasonable match to 50 Ω coaxial cable. Two that are almost painless are as follows:

- Use readily available 75 Ω coax for the λ/4 sections going to each antenna. The λ/4 transformation with 80 Ω at the input will result in 70 Ω for each feed. This will combine to make 35 Ω, resulting in an SWR of 1.4:1. This is usually a workable match. Note that with standard coax, the antennas could be only 0.33 λ apart (perhaps 0.4 λ with foam dielectric coax). This spacing will work almost as well as λ/2, but will not quite cancel the upward lobe and will have a bit less gain.
- Use 50 Ω coax for the lines going to each antenna, but make them an electrical λ/2 long. The λ/2 sections will both repeat the 80 Ω at the input for each feed. These will combine in parallel for 40 Ω, resulting in an SWR of 1.25:1.

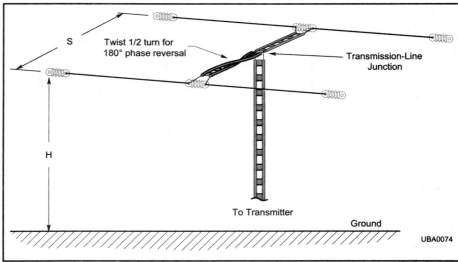

Fig 10-3 — Details of a simple HF end-fire array. Note 180° phase reversal is required at the junction point.

With standard coax this means that the antennas could be up to 0.66 λ apart, a spacing that provides even higher gain.

Another way is sometimes encountered. This one requires some care to result in equal signals going to the two antennas, but does offer a simplicity that is hard to beat. If the two antennas are connected together via an exact λ/2 length of transmission line that is reversed (upper shield to the right element arm, lower shield to the left) the resulting signals to the two antennas will be in-phase. In addition, because the λ/2 transmission line repeats the impedance of the upper antenna the impedance of the upper antenna will be repeated at the junction and the parallel connection should result in an impedance of about 40 Ω.

In a similar way, a 1 λ section of transmission line will also repeat the impedance of the upper antenna at the lower. This time the two antennas will be in-phase if connected to the same side of each antenna. This allows a spacing of up to 0.66 λ, which should result in higher gain and an impedance that is even higher due to mutual impedance and thus a closer match to 50 Ω coax at the junction point.

The 8JK, an Effective and Easy End-Fire Array

The classic HF end-fire array was

described by the late John Kraus, W8JK, Professor Emeritus at Ohio State University and the force behind one of the earliest radio telescope development efforts. The antenna still bears his call sign and is called the "8JK Flat-top Array."

The HF end-fire array is just two dipoles at the same height fed 180° out-of-phase. Skyward energy is cancelled and all available power is radiated towards the horizon. The original 8JK was designed using longer elements, and I'll talk about that configuration again in the multiband antenna chapter. But for now, let's look at the single-band λ/2 resonant case.

The basic configuration is shown in **Fig 10-3**. There is nothing surprising here, but as often the case, the devil is in the details.

Hooking Them Up

A key issue is the mutual coupling of the two closely spaced dipoles. Because the two elements are closely spaced, the mutual inductance between them has a significant effect and reduces the feed-point impedance of each dipole to a very low value. For example, at 0.1 λ, in theory a spacing with high gain, the predicted impedance of each dipole is only around 6 Ω. At 0.15 λ it rises to about 12.5 Ω, perhaps a value that can be matched more efficiently.

If you feed each dipole with a λ/4

length of 50 Ω coax, the impedance at the end of each should be 200 Ω. If we combine them in parallel (remembering to reverse the connections at one dipole to make them out-of-phase) the result should be 100 Ω. Now you either can use 50 Ω coax with an SWR of 2:1, or you can insert a λ/4 section of 75 Ω coax and have an almost perfect match to 50 Ω.

An Easy Collinear Array

The arrays I have described so far in this chapter have been designed to focus their energy at low radiation angles while still maintaining the wide azimuth pattern of a dipole. It is easy to make a horizontal array that doesn't restrict the elevation pattern, but instead has a sharper azimuth pattern to generate gain in the main lobe. This is accomplished by putting two (or more elements along the same axis, or in-line as shown in **Fig 10-5**. This configuration is called a *collinear array*.

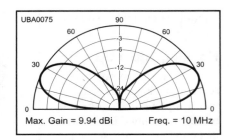

Fig 10-4 — Elevation response of the 10-MHz half-wave end-fire array of Fig 10-3 with 0.15 λ spacing, 50 feet above ground.

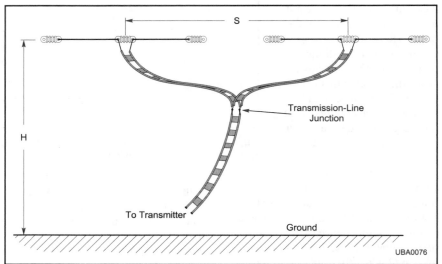

Fig 10-5 — Configuration of two element collinear array.

Changing the Spacing

The minimum center-to-center spacing of λ/2 dipoles is 0.5 λ and the resulting azimuth pattern of a pair of common 30-MHz dipoles, 50 feet high is shown in **Fig 10-6**. The sharpness and gain of the two-lobed pattern increases as the elements are separated, up to a spacing of 0.625 λ, at which point the beamwidth is reduced to 40° and the gain in the main lobes increases to a bit more than 3 dB compared to a dipole (dBd). Above that spacing, the sharpness of the main lobe increases; however, power also appears in additional lobes, reducing the increase in main-lobe gain. This is illustrated in the azimuth plot shown in **Fig 10-8** for a spacing of 1 λ. The multi-lobed pattern can sometimes prove useful if you are trying to communicate with regions at different azimuths at the same time.

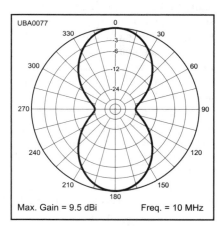

Fig 10-6 — Azimuth pattern of two element collinear array with center-to-center spacing of 0.5 λ.

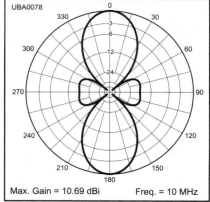

Fig 10-7 — Azimuth pattern of two element collinear array with center-to-center spacing of 0.625 λ.

Feeding Collinear Dipoles

As with other multi-element systems, you must be concerned with the effects of mutual impedance as you try to feed collinear elements. As you might expect, the coupling between wires on the same axis is not very strong and the effects are small except at close spacings. The real part of the impedance of each dipole goes up smartly at $\lambda/2$ spacing, but is negligible for spacings at 0.625 λ and greater. The reactive part of the mutual impedance results in a need to shorten each dipole by about 1% at close spacings in order to achieve resonance. The results are summarized in **Table 10-1**. Each dipole can be fed with equal length sections of 50 Ω line with reasonable (less than 3:1) SWR at 50 Ω expected at the junction. If 93 Ω cable is available, a $\lambda/4$ transformer can be used to transform the impedance above 100 Ω, followed by $\lambda/2$ sections of 50 Ω cable so a pair of paralleled feed lines will be just above 50 Ω. A 75 Ω cable can do almost as well.

Table 10-1

Summary of Collinear Parameters Compared to Single Dipole

C-C Spacing	Gain (dBi)	Gain (dBd)	Beamwidth	Z(Ω)	Length Correction
Single Dipole	7.6	0	85.8°	67	100%
0.5 λ	9.5	1.9	53.4°	88	99%
0.625 λ	10.8	3.2	40°	65	99%
1.0 λ	10.5	2.9	29.8°*	66	100%

*Broadside lobes only.

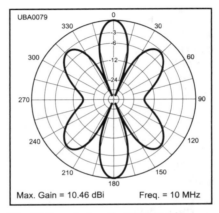

Fig 10-8 — Azimuth pattern of two element collinear array with center-to-center spacing of 1 λ.

The Double Zepp

The extended double Zepp (EDZ), discussed in Chapter 9 to illustrate transmission-line transformation of nonresistive loads, and its companion, the (non-extended) double Zepp, are two easy-to-implement collinear arrays. The double Zepp looks like a center-fed full-wave dipole, but is really two λ/2 dipoles, each end-fed in-phase with a single transmission line. The impedance at that point at resonance is quite high, around 5800 Ω for a 10 MHz system of #12 wire 50 feet high.

Using our expression from Chapter 9 for finding the impedance of a λ/4 matching section to match to 50 Ω, reveals that a 538 Ω line is needed. While you could construct that line, a common 600 Ω line will transform the impedance to 62 Ω, yielding a usually acceptable SWR of 1.2:1 for 50 Ω line. The configuration was shown in **Fig 10-9**.

The azimuth pattern of the double Zepp is the same as a dual-fed collinear with 0.5 λ spacing. The extended-double Zepp with its matching section is described in Chapter 9. The extended double Zepp azimuth pattern is shown in Fig 9-3. Note that the out-of-phase radiation from the center section of the extended double Zepp results in a slightly ragged pattern compared to the 0.625 λ dual-fed collinear; however, it is close enough to work well in most applications and is often easier to construct.

Notes

[1]J. Kraus, W8JK (SK), "Directional Antennas with Closely-Spaced Elements," *QST*, Jan 1938, pp 21-25.

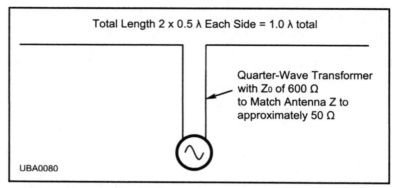

Total Length 2 x 0.5 λ Each Side = 1.0 λ total

Quarter-Wave Transformer with Z_0 of 600 Ω to Match Antenna Z to approximately 50 Ω

UBA0080

Fig 10-9 — Configuration of center fed Zepp version of two element collinear array.

Review Questions

10-1. Discuss the benefits and limitations of 2-element broadside arrays as a means of focusing radiation at low elevation angles.

10-2. Discuss the benefits and limitations of 2-element end-fire arrays.

10-3. Consider the collinear array of Fig 10-5. Why can't it be fed via a pair of λ/4 matching sections?

10-4. What factors might you consider to decide between using an extended-double-Zepp and a dual 0.625-λ spaced dual-fed collinear pair?

Chapter 11

Wideband Dipole Antennas

The X-shaped spreaders on either side of the center of the wideband cage dipole at W1AW

Contents

What's the Story About Bandwidth?

To date I have been discussing antennas designed for a single frequency, usually a convenient 10 MHz. It is important to recognize that any signal, except for a steady carrier, includes sidebands around the operating frequency. To avoid distortion of the transmitted signal, the antenna needs, at a minimum, to be able to radiate all the frequencies that are part of the transmitted signal — without distortion.

The Effects of Bandwidth Restriction

All of the antenna parameters we have discussed — SWR, elevation pattern and azimuth pattern — will change when the transmit frequency is changed. Each can also have the effect of making a change in the received signal level, even if the receiving antenna has no restrictions itself.

For narrowband signals, usually the first transmit parameter to exhibit a noticeable difference is the SWR. While the SWR itself won't result in a change in antenna response, the resulting change in impedance as seen at the transmitter will often result in a change in the power delivered to the load. Depending on the loss of the transmission line to start with, an increase in SWR can also result in an increase in line loss, also reducing the power delivered to the antenna.

How Much Bandwidth do You Need?

Different transmission modes require different bandwidths to carry the associated sidebands. **Table 11-1** is a summary of common signal types encountered in radio systems, along with their requisite bandwidths. These bandwidths apply to a transmission system operating on a single carrier frequency. While there are many such single-frequency broadcasters and communications systems, there are also many that require operation on multiple frequencies.

Table 11-1

Bandwidths of Common Transmission Modes

Mode	Use	Typical Bandwidth
Radiotelegraph	Manual Keying	<100 Hz
Radioteletype	Manual Keyboard	<300 Hz
SSB Voice	Communication	3 kHz
AM Voice	Communication	6.6 kHz
AM Voice/Music	Broadcast	10 kHz
FM Voice	Communication	15 kHz
FM Stereo Music	Broadcast	150 kHz
Analog Television	Broadcast	6 MHz
Pulse	Data/Radiolocation	2/Pulse Width[1]

[1]Typical, varies significantly with risetime and pulse shape.

These fall in a few categories that I'll discuss below.

Operation Across a Band of Frequencies

Instead of a single frequency of operation, some services require users to be able to operate on different channels within a band of frequencies. Examples of this kind of operation include the Amateur Radio service that allows operation on multiple, fairly wide bands across the radio spectrum. Another example is the maritime radio service.

Maritime HF operation occurs in multiple bands across HF and into VHF to allow for different ranges and propagation conditions. Within each band the operator initiates communication on a *calling channel* and then switches to one of a number of *working channels* to pass traffic. The selected working channel will depend on the type of service, category of distant station (shipboard, shore station, drawbridge or USCG, as examples — each with different assigned groups of channels) and whether or not a particular channel is in use, so most stations must be prepared to operate on virtually any channel.

In either of the above examples, operational simplicity results if the design of the antenna is such that an operator can select any channel within a band without having to make a change in the antenna system or the transmitter output circuit. The ability to accomplish this is usually a function of the width of the band as a fraction of the band's center frequency.

Do We Have a Problem?

A good example of such a band is the US 80 meter amateur band. It extends from 3.5 to 4.0 MHz, a width of 0.5 MHz, or 13.3% of the band center. Other amateur HF bands have a narrower percentage bandwidth, so this is a good limiting case to look at. If you can make 80 meters happen, you can succeed on any other HF amateur band.

I'll start examining the issue by looking at a good baseline configuration — a $\lambda/2$ long thin wire dipole. I will assume that you have a radio system that can work into a 2:1 SWR at 50 Ω. I will use a standard dipole height of 50 feet, and find a length that makes the SWR approximately equal at the band edges. The result is shown in **Fig 11-1**. The 2:1 SWR bandwidth is around 150 kHz, with an SWR that rises above 5:1 at the band edges. It looks like something will have to happen to obtain an antenna that meets the requirements — it would take more than three separate dipoles of this simple design to cover the entire band! At around 130 feet in length, this is not a solution for everyone — besides, I don't want to require the operator to have to remember to switch antennas in mid-operation as channels are changed.

So What Should You Do?

The first step is to observe that I have posed a bit of a hurdle by not having an antenna that has an impedance of 50 Ω at any frequency in the band. Assume that I can design a wideband 69:50 Ω transformer that will transform the antenna mid-band load of 69 Ω to the 50 Ω that the transmission line and radio want to see. There are a number of ways to do that, so please assume I have one at hand. The result is shown in the plot of **Fig 11-2**. Note that I now can cover 200 kHz with an SWR of 2:1 — I'm headed in the right direction but I'm not there yet.

In Chapter 4, I discussed the relationship of antenna length-to-thickness ratio as it affects impedance change with frequency. Another term for that is *bandwidth*, so let's explore that as a possibility. Using *EZNEC*, it is easy to adjust the wire diameter to get the SWR bandwidth I want. I find that by increasing the wire from #12 to 24 inches in diameter I can achieve the desired SWR bandwidth, as shown in **Fig 11-3**.

Using tubing for elements is quite common at higher frequencies, and thick-diameter tubing is found as a material in a number of broadband VHF antenna designs. Unfortunately, even with the shortened length of the thicker antenna (from 127 to 118 feet), hoisting that much thick tubing up 50 feet may not be a feasible solution, in spite of how easy it is to model.

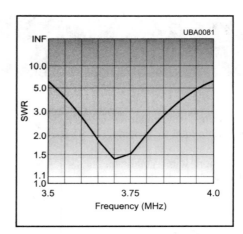

Fig 11-1 — SWR plot of 80 meter dipole of #12 wire at 50 feet.

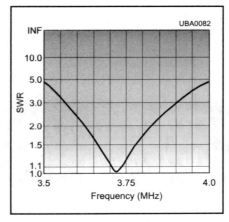

Fig 11-2 — SWR plot of 80 meter dipole of #12 wire at 50 feet referenced to 69 Ω.

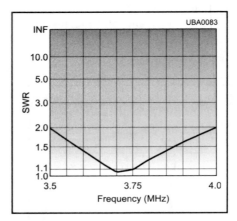

Fig 11-3 — SWR plot of 24 inch diameter 80 meter dipole at 50 feet referenced to 69 Ω.

Enter the Cage Antenna

Instead of using tubing, you can simulate the current flow on a tube by providing multiple wires along its length at some radius from the center. The more wires you have, the closer it will act like a piece of tubing. Consider an antenna made of four wires, equally spaced on the corners of a square with an 8 foot diagonal shown in **Fig 11-4**, you achieve virtually the same SWR performance as the 2 foot diameter tube, with much less weight. Additional wires will result in a smaller cross section required, but a heavier structure. A cage antenna of six wires about 1 foot across was used at the US end of the first amateur transatlantic communication back in 1923.

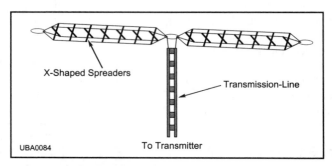

Fig 11-4 — Details of a four wire cage antenna.

A Two-Dimensional Cage — the "Fat" Dipole Antenna

Instead of making a three dimensional skeleton of tubing, you can simplify things even further, by adding a parallel wire some distance from the first. The configuration is shown in **Fig 11-5**. The outside ends are both at the same potential, so they can be connected, or not as you wish. Wider spacing will determine the bandwidth for this band, but a spacing larger than about 1 foot doesn't change things much. I have made an *EZNEC* SWR plot of such a dipole with a 1 foot spacing, as shown in **Fig 11-6**. Compare this with the SWR plot of a single wire "thin" dipole in Fig 11-1.

While this configuration almost meets a 3:1 SWR goal, rather than my 2:1 SWR goal for 80 meters, when scaled it will cover any higher HF amateur band. Some radios, more tolerant of SWR, will work across all of 80 meters as well.

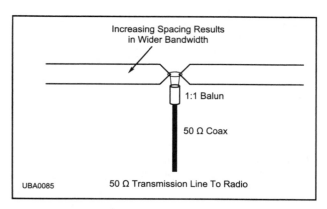

Fig 11-5 — Details of an HF two dimensional "fat dipole" configuration.

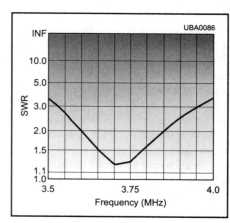

Fig 11-6 — SWR plot of an 80 meter fat dipole with 1 foot spacing between wires.

The Conical Dipole and Fan Dipole

An alternative to a large-diameter cylinder antenna, or its wire equivalent, is a dipole made from conical sections with their apexes at the center of the antenna. Again, metal cones are feasible at VHF and above, but mechanically difficult at HF. An approximation to a cone can be constructed using multiple wires on each side of the antenna center. A five wire configuration is shown in **Figs 11-7**

and **11-8**. The SWR plot in Fig 11-8 assumes the use of a transformer to match from 40 to 50 Ω. Again, this doesn't quite meet my 2:1 criterion across the band, but it is pretty close and will work for bands with a band-to-center frequency ratio of about 9%.

By using a two dimensional equivalent of a cone, you have the *fan dipole* shown in **Fig 11-9**. This provides considerable mechanical

simplicity, since a single halyard can easily hoist each end, perhaps guided by a spacer or two. I have selected a spacing of 4 feet at the ends for the SWR curve shown in **Fig 11-10**. As you would expect, it is not quite as effective as the three-dimensional version, but it may be sufficient for many applications.

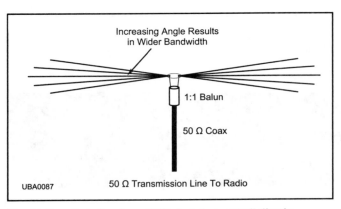

Fig 11-7 — Details of an HF "skeleton cone" dipole configuration.

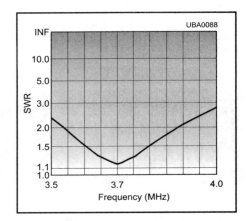

Fig 11-8 — SWR plot of an 80 meter skeleton cone dipole with 4 foot radius cone.

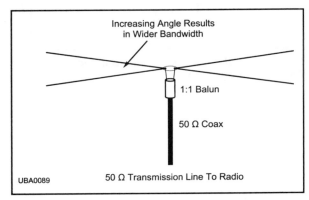

Fig 11-9 — Details of an HF two dimensional cone or "fan dipole" configuration.

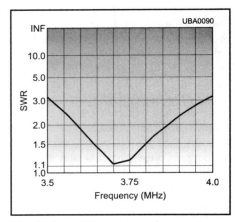

Fig 11-10— SWR of an 80 meter fan dipole configuration with end spacing of 4 feet.

The Folded Dipole Antenna

An interesting dipole variant is one called a *folded dipole*. This is actually two dipoles — one fed in the center — and one fed at the ends, in close proximity to each other, as shown in **Fig 11-11**. The combination of the two antennas results in two differences from the usual dipole:

The impedance is four times the impedance of a single dipole — nominally 280 Ω rather than 70 Ω — a close match to common 300 Ω transmission line. The bandwidth is similar to a dipole with wires of the same spacing; however, the folded dipole offers an opportunity to have such a dipole with a higher impedance — especially beneficial in some kinds of arrays.

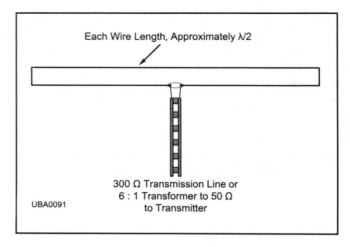

Fig 11-11 — Details of an HF folded dipole configuration.

The Terminated Wideband Folded Dipole Antenna

It will likely not come as a surprise to anyone who has read this far that one way to have a wideband load on a transmitter is to connect it to a resistor. Unfortunately, while providing a good match, resistors don't generally radiate energy, except perhaps as heat, very effectively!

An interesting combination of a resistive load and long connecting wires that does divide power between resistive heating and radiation from the antenna is called the *terminated wideband folded dipole* (TWFD). The configuration is shown in **Fig 11-12**. They can be made in different lengths; however, I have selected a 95-foot version for modeling. This antenna is available as a commercial product, advertised to cover the range of 3 to 30 MHz with a reasonable SWR. One look at the plot in **Fig 11-13** indicates that it appears to deliver on the bandwidth promise.

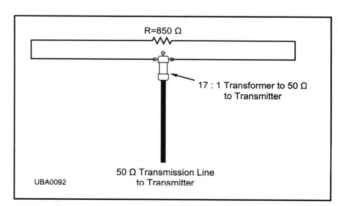

Fig 11-12 — Details of a terminated wideband folded dipole (TWFD) configuration.

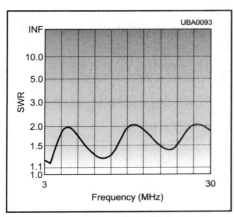

Fig 11-13 — SWR of 95 foot terminated wideband folded dipole (TWFD).

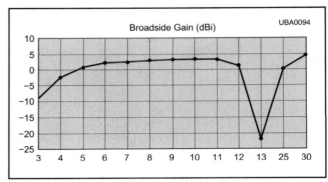

Fig 11-14 — Broadside gain of 95 foot terminated wideband folded dipole (TWFD).

Not surprisingly, there is a penalty associated with this type of antenna and that is in radiating efficiency. Any power that is dissipated in the resistor is not radiated and thus there is less available than would be the case of an antenna made of conducting material. **Fig 11-14** shows the broadside gain as a function of frequency. Compare these values to a λ/2 dipole over typical ground with a broadside gain of at least 7 dBi (increasing somewhat with frequency), and you will see that there is a penalty of at least 4 dB starting above its λ/2 frequency of around 5 MHz. Below that its penalty is significantly higher, reaching about 16 dB at 3 MHz.

The dip in broadside output at 20 MHz is due to pattern splitting.

The antenna actually has a gain of around 4 dBi at 50° off broadside. This may or may not be an issue depending on how it is used. If the far end destination is the same at all frequencies, then this broadside null would make the antenna a poor choice if it always has to radiate towards the broadside direction — such as in a point-to-point automatic link enablement (ALE) system, which will be searching for frequencies that have good propagation and automatically switch equipment to an optimum frequency.

In spite of the fact that many dedicated antenna fans don't like to see resistors in antennas, the TWFD may be a good choice in some applications. At the lower portion of the HF spectrum, received signal-to-noise ratio (SNR) is often limited by external noise, so a loss in the antenna system won't reduce received SNR. On the transmit side, a government-funded station may be able to just throw a switch to add 10 dB of power and more than make up for the antenna loss at many frequencies! Being able to change frequencies without worrying about antenna operation is a great convenience, especially for untrained or stressed operators.

Notes

[1]*TLW* is supplied with *The ARRL Antenna Book*, 21st Edition, available from the ARRL Bookstore at **www.arrl.org/ catalog/** order number 9876 — $44.95.

Review Questions

11-1. Describe some applications for which a thin-wire dipole would have sufficient bandwidth.

11-2. What would be the likely effects of operating a 3.5 MHz transmitter into an antenna with an SWR as shown in Fig 11-1?

11-3. Under what conditions might the TWFD antenna be a good choice? What are the penalties of using one compared to separate dipoles for a number of channels?

Chapter 12

Multiband Dipole Antennas

Resonant circuits allow both these antennas to operate on multiple bands.

Contents

What Do We Mean by Bands?

In this context, a band is a range of frequencies assigned to a particular function. You are familiar with the MF AM broadcast band, covering 550 to 1700 kHz, a range that is a challenge for wideband antennas. In addition, there is broadcasting throughout the HF region in multiple bands as shown in **Table 12-1**. Other services with multiple HF bands include the Amateur Radio, maritime and aeronautical services.

In the case of services using HF allocations, it is common to have multiple bands assigned to the same function, in some cases carrying the same information, so that as the ionospheric conditions change a listener can shift from a channel in one band to a channel in another and still maintain communication. In some cases, as with a diversity receiving system using frequency diversity, this happens automatically. In other cases a manual frequency change is required. It is convenient to be able to use the same antenna to cover multiple bands. While having a separate antenna for each band is a possibility, it often is not practical from a real-estate resource standpoint.

Table 12-1

Selected US HF Frequency Allocations by Service* (MHz)

Amateur	Aeronautical	Maritime	Broadcast
		3.0 – 3.155	
			3.2 – 3.4**
	3.4 – 3.5		
3.5 – 4.0			
			3.9 – 4.05**
		4.0 – 4.438	
	4.65 – 4.75		
			4.75 – 4.995**
			5.005 – 5.1**
5.33 – 5.41***			
	5.45 – 5.73		
			5.9 – 6.2
		6.2 — 6.525	
	6.525 – 6.765		
7.0 – 7.3			
			7.2 – 7.35**
		8.195– 8.815	
	8.815 – 9.040		
			9.5 – 9.9
	10.005 – 10.1		
10.1 – 10.15			
	11.175 – 11.4		
			11.6 – 12.1**
		12.23 – 13.2	
	13.2 – 13.36		
			13.57 – 13.87
14.0 – 14.35			
			15.1 – 15.6
		16.36 –17.41	
			17.9 – 18.03
18.068 – 18.168			
		19.68 – 19.8	
21.0 – 21.45			
			21.45 – 21.85
	21.924 – 22.0		
24.89 – 24.99			
		25.07 – 25.21	
			25.67 – 26.1
		26.1 — 26.175	
28.0 – 29.7			

*This should be considered representative and not official.
There are some variations between allocations in different regions of the world. Allocations do not necessarily imply usage.
**Outside US only.
***Five specific channels allocated in this range.

What Does Multiband Mean?

In our last chapter, we discussed *wideband* dipole antennas. It should be clear that if multiple bands are within the bandwidth of a wideband antenna, it could also serve as a *multiband* antenna. What I will be talking about in this chapter is a group of dipole antennas types that can operate on multiple distinct bands — but not necessarily all the frequencies in between, as would be the case with a wideband antenna.

Taking Advantage of Multiple Resonances

As noted in Chapter 4, particularly in Figs 4-3 through 4-5, at frequencies that are approximately odd multiples of a dipole's λ/2 frequency a dipole will show an impedance very similar to that at its λ/2 frequency. If you have allocations that include bands that are odd multiples of the frequency of other bands, you can take advantage of this to get an additional band for "free." As noted in Table 12-1, most services have band with, or close to, such a frequency relationship.

As an example, consider the US amateur bands at 7 and 21 MHz (40 and 15 meters). If you construct a λ/2 dipole for 7.15 MHz, at the usual height of 50 feet above ground, you will find that it has a 50 Ω SWR curve such as that shown in **Fig 12-1**. While the SWR is very close to the usual design goal of 2:1 across the band, by inserting a 1.5:1 transformer at the feed it is even better, as shown in **Fig 12-2**.

Matching a Dipole on Two Bands

Now try feeding the antenna across the 15-meter band. You would think this should work just fine. Oops — as shown in **Fig 12-3**, things don't quite work that way. On 15 meters the same physical height above ground is three times as high in terms of wavelength than on 40 meters. And other factors have similar effects. The result is that the antenna is mistuned on the higher band. By retuning the antenna for 15 meters in the *EZNEC* model, shortening it by 1.4 feet,

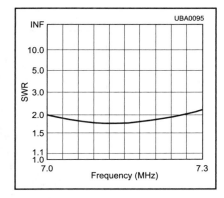

Fig 12-1 — SWR plot of λ/2 dipole for 7.15 MHz, at a height of 50 feet above ground.

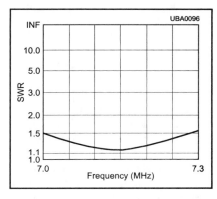

Fig 12-2 — SWR plot of antenna in Fig 12-1 with 1.5:1 transformer at feed point.

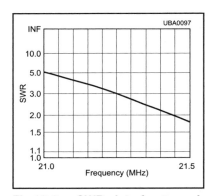

Fig 12-3 — SWR plot of antenna in Fig 12-2 over the 15 meter band.

you end up with the result shown in **Fig 12-4**. Note that this assumes that the 1.5:1 transformer is designed to work over the 7 to 21.45-MHz range — not usually a problem.

Now what is happening on our original band — 40 meters? As shown in **Fig 12-5**, not surprisingly, it is not quite as good as it was before, but it comes very close to meeting the

design criterion. It is generally true that the tuning will be much more critical on the higher band, so tuning there — perhaps with a bit of "Kentucky windage" back and forth — it is often possible to come up with an antenna with a satisfactory match on both bands.

Is That All There is to it?

It's fine to be able to properly feed energy into — and receive energy from — an antenna on two bands, but you also need to consider where the energy goes to or comes from. Your antenna at a height of 50 feet will be about 0.36 λ high on 40 meters and 1.08 λ high on 15 meters. This difference in electrical height means that the peak of the elevation lobe will move from 40° on 40 meters to 13° on 15 meters. This will have implications for your transmission range, because the 40° elevation is optimum for medium ranges (say, out to 2000 miles) while a 13° takeoff angle will be best at longer ranges — it all depends on the effective height of the ionosphere.

In addition, the azimuth pattern will change from the familiar bidirectional "figure-eight" pattern of a dipole to a much more complex pattern, as shown in **Fig 12-6**. This is a characteristic of center-fed antennas that are used on frequencies above 1 λ long. Note that, just as in the case of elevation angles, this may be either a benefit or liability depending on where the other end of the communication link is located. In this case, the gain at the peak of the broadside beam is almost identical for the two cases, although more precise aiming accuracy is required on the higher band.

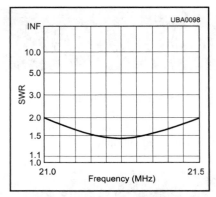

Fig 12-4 — SWR plot of antenna in Fig 12-2 shortened by 1.4 feet over the 15 meter band.

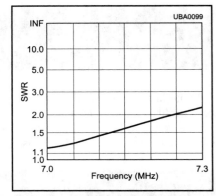

Fig 12-5 — SWR plot of antenna of shortened antenna over the 40 meter band.

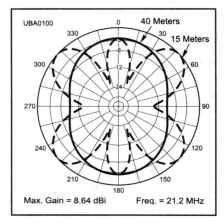

Fig 12-6 — Azimuth plot of antenna on 40 meters (solid line) compared to 15 meters (dashed line).

Multiple Resonator Antennas

The use of antennas on odd harmonic resonances is convenient, but will likely never cover all your multiband antenna requirements. There are a number of approaches that can provide similar performance, usually at some cost in terms of additional complexity.

Parallel Dipoles for Multiple Bands

One approach is to join multiple antennas together in the same space and using the same transmission line. This is generally referred to as *parallel dipoles*, and is shown in **Fig 12-7**.

The concept on which this is based is that with antennas at widely different frequencies, each will be resonant within its band and will have a sufficiently high impedance on other bands so it doesn't readily accept power from the transmission line. While this makes a fine story, and the arrangement is often successfully employed, there are two pitfalls that can limit the usefulness of this approach:

• In addition to the connected parallel impedance of the additional dipole, there is a significant mutually coupled impedance that may be hard to predict. It will range from negligible to significant as the direction of the second dipole is changed from being perpendicular to, towards being parallel to the first.

• While the impedance may be high, except for harmonic frequencies, it will generally be reactive resulting in a need to retune the antenna lengths compared to a single-band dipole.

You must do some trimming to make everything work. The good news is that any energy that goes into the off-band dipole will be radiated, although not necessarily in the desired direction.

Parasitically Coupled Multiband Antennas

A solution that is somewhat similar to the previous parallel dipoles is actually another form of parallel dipole — in a different sense of the word. While parallel in the previous section referred to a parallel electrical connection, it is also possible to couple from one parallel dipole to another tuned to a different frequency without direct connections. This is based on mutual impedances and is called *parasitic coupling*. See **Fig 12-8**.

Several commercial manufacturers of Amateur Radio directive antennas for the higher HF regions have used this concept successfully. While the concept is simple, successful implementation may be a bit tricky. Modeling efforts indicate that spacing between driven and parasitic element is critical

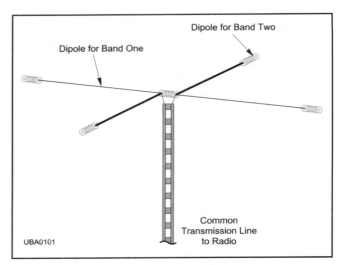

Fig 12-7 — Configuration of parallel dipoles for two bands.

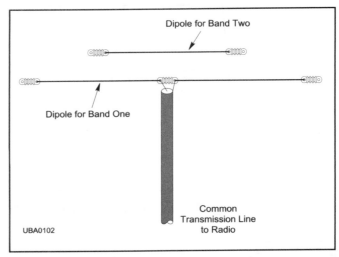

Fig 12-8 — Configuration of parasitically coupled dipoles for two bands.

for successful operation. The spacing between driven (lower frequency) dipole and parasitically coupled element needs to be about 0.003 to 0.004 λ to achieve appropriate coupling.

The good news is that the coupled element doesn't have much impact on the tuning or pattern performance of the driven (lower frequency) dipole. The bad news is that the SWR bandwidth of the higher frequency element(s) is narrower than would be the case for a connected dipole on the same band. Adding additional bands in this way makes for additional interaction and complexity, although the promise of a compact and inexpensive multiband structure certainly seems attractive.

Antennas with Resonant Traps

A parallel-resonant circuit has the property that it has a high impedance at its resonant frequency. A high impedance acts very much like an insulator. By placing such a circuit on each side of a dipole antenna at the $\lambda/2$ end points corresponding to the resonant frequency, the antenna effectively ends there for that frequency. At a lower frequency, the parallel-resonant circuit is no longer resonant and looks like an inductance. The configuration is shown in **Fig 12-9**.

Fig 12-10 shows the SWR plot of such an antenna made for the 80 and 40 meter amateur bands. The traps are made from 10 µH inductors in parallel with 50 pF capacitors, along with the additional capacitance of the leads. This results in resonance around 7 MHz. On the lower frequency, the trap looks like an inductance, and it serves to act as if part of the antenna became rolled up, resulting in a shortened overall electrical length.

There is nothing to prevent the addition of other traps on each side to provide operation on additional bands, and you will frequently encounter this. An advantage of this arrangement over others to be discussed is that this method maintains the broadside response pattern of a dipole on each band — not the case with some other multiband designs.

On any higher-frequency bands without traps, the trap acts like a capacitor and thus tends to make the antenna look electrically shorter than its physical length. Since there are an infinite number of combinations of L and C that can be resonant at any frequency, it is possible to select a pair that are resonant at 7.15 MHz and take advantage of this property. Chester Buchanan, W3DZZ, developed such a trap dipole in the 1950s with resonances in the 80, 40, 20, 15 and 10 meter bands — all the HF amateur bands at that time.[2] **Fig 12-11** shows the *EZNEC* SWR plot of such an antenna.

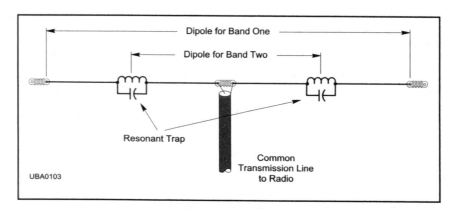

Fig 12-9 — Configuration of two-band trap dipole for 80 and 40 meters.

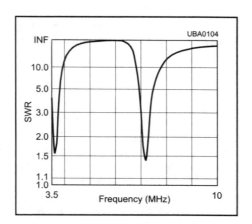

Fig 12-10 — SWR of two-band trap dipole in Fig 12-9 showing resonances on 80 and 40 meters.

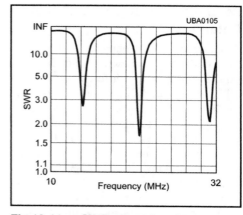

Fig 12-11 — SWR of two-band trap dipole in Fig 12-9 from 10 to 32 MHz showing additional resonances on 20, 15 and 10 meters.

The Center Fed Zepp

As noted in **Table 4-1** and Figs 4-3 to 4-5 and earlier sections of this chapter, you can use dipoles of electrical lengths other than λ/2 effectively. The consequence of using an antenna that isn't resonant (or which is resonant but doesn't match the transmission feed line) is that you will need a matching network to couple power from the transmitter to the antenna system. If the matching system, generally called an *antenna tuner*, is located at the antenna itself, 50 Ω coaxial cable can be used for the run to the radio room. If the matching network is located near the radio equipment, particular attention must be paid to losses in the transmission line, since the SWR will be quite high on many bands.

A popular antenna system for wide frequency coverage is one called a *center fed Zepp*. This antenna is a variation of the antenna that used to trail behind Zeppelin airships in the 1930s. The antenna is not "wideband" in the same sense as previous examples. Instead, it has a feed impedance that varies widely with frequency, but which is compensated for at the radio end of the transmission line by a tunable matching network. The center fed Zepp can operate on all frequencies above (and even somewhat below) its λ/2 resonant frequency.

The concept of this antenna is to start with a λ/2 dipole at the lowest frequency of operation, as shown in **Fig 12-12**. Feed the dipole with a low-loss transmission line and use a matching device at the radio end to adjust for whatever impedance is found on the frequency you wish to operate.

The design parameters for such an antenna are based on λ/2 at 3.6 MHz and fed with 100 feet of nominal 450 Ω (actual Z_0 around 400 Ω) "window" transmission line. The resulting SWR curve is shown in **Fig 12-13**. The resulting line losses are shown in **Table 12-2**. The dipole length in feet is found by dividing 468 by 3.6, resulting in 130 feet. I will assume a height of 50 feet, reasonable for many locations with mature trees.

Note that in spite of an SWR as high as 12:1, the worst-case loss in this low-loss transmission line is only around 0.5 dB — hardly noticeable. By contrast, consider feeding the antenna with typical 50 Ω coax, which would be fine at the antenna's λ/2

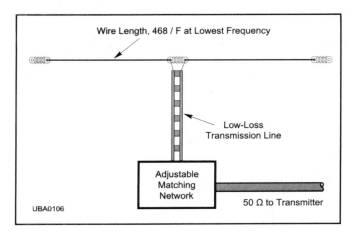

Fig 12-12 — Configuration of center fed Zepp.

Table 12-2

Design Parameters for Center-Fed Zepp

Frequency	Antenna Z, Ω	SWR at 400 Ω	Line Loss (dB)
3.6	63 − j 39	6.4	0.128
5.3	351 + j 868	7.25	0.176
7.2	4253 − j 1232	11.5	0.399
10.1	105 − j 595	12.4	0.501
14.2	1472 + j 1816	9.45	0.428
21.2	648 + j 1042	6.27	0.385
28.3	348.4 + j 606	4.43	0.318

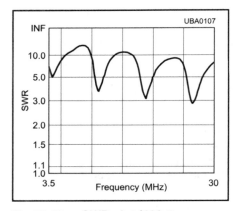

Fig 12-13 — SWR plot (400 Ω reference) of center-fed Zepp.

resonance. If you used lightweight RG-58A cable, for example, the SWR at 28.3 MHz would be 28.2:1. And the total loss in the cable would be a staggering 10.5 dB — burning up more than 90% of the power in the transmission line, with less than 10% of the power actually delivered to the antenna.

Note the azimuth radiation patterns for selected bands shown in **Fig 12-14** through **Fig 12-17**. At 7.2 MHz, where the antenna is about 1 λ long, the pattern looks very much like what you would expect to see from a λ/2 dipole. The sharp-eyed observer will note that compared to that of a λ/2 dipole, the lobes are sharper and the gain is a bit higher. As the electrical length moves above 1 λ, the pattern gets significantly more complex. This can be a mixed blessing. On the plus side, you get more gain and coverage into different regions on the higher bands. On the other side, if you wish to communicate (or listen) to specific locations, you may wish you had the dipole's simple pattern on every band.

Another band-to-band difference is that at the single height of 50 feet, the elevation angle response is lower as the operating frequency is increased. This allows for longer-range communication at higher frequencies, again a possible benefit, depending on requirements.

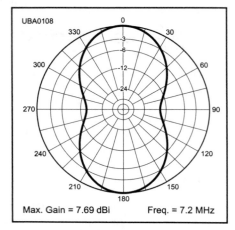

Fig 12-14 — Azimuth radiation pattern of 3.6 MHz center fed Zepp at 7.2 MHz.

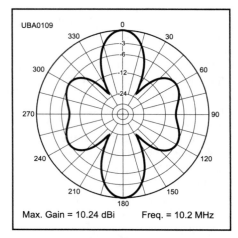

Fig 12-15 — Azimuth radiation pattern of 3.6 MHz center fed Zepp at 10.1 MHz.

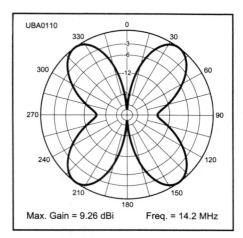

Fig 12-16 — Azimuth radiation pattern of 3.6 MHz center fed Zepp at 14.2 MHz.

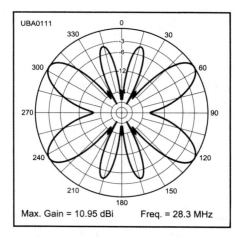

Fig 12-17 — Azimuth radiation pattern of 3.6 MHz center fed Zepp at 28.3 MHz.

The G5RV Dipole

The G5RV, named for its late developer Louis Varney, G5RV, is a variation of the center fed Zepp. Varney originally designed the dipole to be 1.5 λ long on the 20 meter band, where it could achieve a modest amount of gain over a half wave dipole, and yet could also operate on other amateur bands with an antenna tuner. This approach is in contrast to most multiband single-wire dipoles described earlier, which start off with a half-wave-long wire at the *lowest* frequency of operation. The G5RV employs a simple feed system using a transition between open-wire and coaxial transmission lines and has moderate SWR on bands other than 20 meters.

The G5RV antenna has a "flat top" length of 102 feet and Varney fed his with what he called a "20 meter matching section" of 34 feet of low-loss 300 Ω TV-ribbon line, followed by enough 75 Ω coax to get down to the transmitter. Most designs nowa-days use more commonly available 450 Ω window line for the 34 foot matching section.

A plot of the SWR of an *EZNEC* model is shown in **Fig 12-18** for 50 Ω coax. On 15 meters the SWR is rather high, in excess of 25:1, for example. In the 20-meter band, the SWR rises as high as about 4:1.

I have used a few G5RVs over the years and believe that their simplicity, more than their performance, accounts for their popularity. In my experience it is very difficult to adjust the two available parameters — flat-top length and matching line length — and end up with good SWR performance on many amateur bands. That being said, I have achieved a 5:1 SWR on most bands and find that is low enough to use low loss 50 Ω coax if the runs are short enough. As Varney recommended, a tuner makes life easier at the radio end of the feed line.

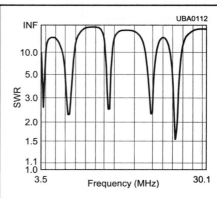

Fig 12-18 — SWR plot (50 Ω reference) of G5RV dipole.

Wrap-Up

It is clearly possible to use a dipole antenna or its variants to achieve satisfactory operation on multiple bands. This can be very helpful to those trying to install an HF station on a property of limited size. In addition, the ability to not have to switch antennas when changing frequencies is a major benefit.

The downside is that while a single-band dipole is very easy to tune and place into operation, most multiband antennas (the center-fed Zepp is a notable exception) require considerable interrelated tuning steps to get everything working as anticipated.

There are other types of multiband antennas besides those based on the λ/2 horizontal dipole. I will discuss them in subsequent chapters.

Notes

[1] *TLW* is supplied with *The ARRL Antenna Book*, 21st Edition, available from the ARRL Bookstore at **www.arrl.org/catalog/** order number 9876 — $39.95.

[2] C. Buchanan, W3DZZ, "The Multimatch Antenna System," *QST*, Mar 1955, pp 22-24.

Review Questions

12-1. Why is a multiband antenna a benefit?

12-2. Of the antennas described, which provides maximum signal strength in the broadside direction on all bands?

12-3. What are the benefits of a multiband antenna with different azimuth patterns on different bands?

Chapter 13

Vertical Monopole Antennas

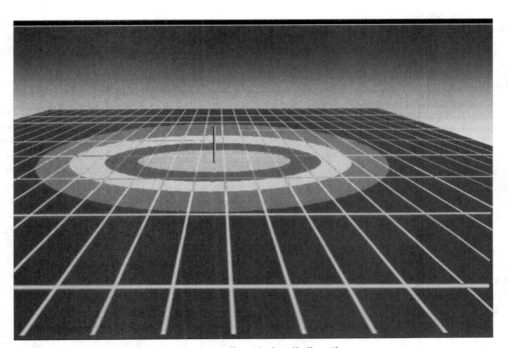

The field from a monopole extends uniformly in all directions.

Contents

How Can Half a Dipole Work?

If, by just looking at the name, you conclude that a *monopole* ought to be half a **di**pole — you'd be right! The hard part comes in understanding what happens to the *other half* of the antenna. The key is that a monopole is located adjacent to some kind of ground. How well it works is closely related to the nature of the nearby — as well as more distant — ground.

The key to the operation of a λ/4 vertical is that the electric field generated by a source driving the antenna terminates on the ground instead of on the other half of a dipole. **Fig 13-1** compares the electric field of a vertically oriented λ/2 dipole in free space with the field of a λ/4 monopole above a perfectly conducting ground. Note that the fields are identical in the region above ground. Not surprisingly, they don't exist below ground.

Radiation from a Monopole

The radiation elevation pattern of a λ/4 monopole over perfect ground is shown in **Fig 13-2**. With a "perfect ground" both under the antenna and at some distance, the reflected and direct signals combine to result in maximum radiation at the horizon, just as you would have from a vertical dipole in free space. This condition is similar to what occurs with a vertical monopole erected in an ocean communicating to another vertical antenna in the same ocean at distances at which the Earth is approximately flat. For longer distances, the broadside signal will maintain its direction tangent to the Earth's surface, resulting in long distance propagation if the ionosphere supports that frequency.

The Effects of Imperfect Ground

In many cases, an ocean is not at hand and the reality of imperfect

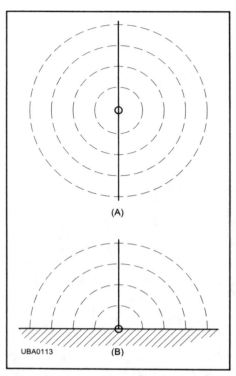

Fig 13-1 — At A, electric field lines of a vertical dipole. At B, monopole over perfect ground.

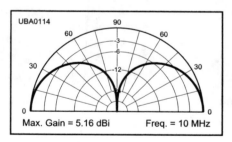

Fig 13-2 — Elevation pattern of antenna in Fig 13-1B.

ground must be taken into account. The properties of real ground that change the way an antenna performs can be characterized by its conductivity and dielectric constant. These vary widely by region, as well as by localized conditions within a region. The ground has two distinct effects on the performance of a monopole. Many people confuse these, so it is important that they be clearly understood.

The Ground as a Part of the Antenna Load

To connect a transmission line (usually a coaxial cable) to a monopole, you have two connections to make. The center conductor is usually connected to the insulated monopole. The shield must be connected to something else — *ground*. This ground connection must carry every bit of current that the monopole does, in order for the antenna to accept the transmitter power. A major hitch is that imperfect ground doesn't have a good GROUND terminal to connect to.

To properly connect to the ground, we need to connect to the region on which the electric lines of force terminate. This generally requires bare radial wires at or beneath the surface of the ground. Note that the "ground rods" often used for lightning protection and power neutral grounding connect largely below the surface. Because of the skin effect, higher frequency currents tend to flow near the surface of real soil. The resulting connection will have some effective resistance, lots in the typical ground rod, less as more and longer ground radials are used near the surface. For the purpose of determining the power that actually reaches the antenna, the situation can be analyzed as an equivalent circuit. See **Fig 13-3**.

At resonance, $L_A = C_A$ and thus the resistance shown as R_R represents the load of the antenna itself, as if it were over perfect ground. This is called the *radiation resistance* of the antenna. Power dissipated in the radiation resistance becomes the actual power radiated by the antenna. The resonant radiation resistance of this ideal monopole is just half of that of a resonant dipole in free space, or 36 Ω.

R_G represents the equivalent RF resistance of the ground connection.

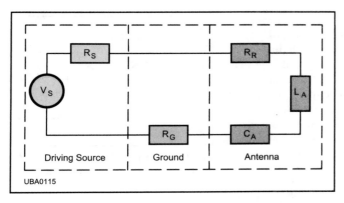

Fig 13-3 — Equivalent electrical circuit of monopole antenna over real ground.

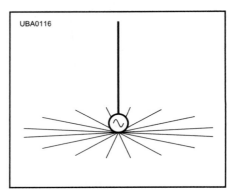

Fig 13-4 — Configuration of ground-mounted monopole with radial ground system.

Power dissipated in this resistance ends up warming up the ground — not usually a goal of the antenna designer. The total power delivered by the source is thus split up into a portion that is radiated and a portion that is dissipated in the ground. The efficiency of a monopole can be defined as the fraction of the power delivered by the source that ends up being radiated by the antenna. For a current i that flows in the circuit, the power delivered is:

$$P = i^2 \times (R_R + R_G) \qquad \text{Eq 13-1}$$

While the radiated power is just:

$$P_R = i^2 \times R_R \qquad \text{Eq 13-2}$$

Thus the resulting efficiency is:

$$P_R / P_R = R_R / (R_R + R_G) \qquad \text{Eq 13-3}$$

Achieving High Efficiency with a Ground Mounted Monopole

The usual ground system for feeding power to a vertical monopole consists of buried radial wires extending out in all directions from a point just beneath the base of the antenna. See **Fig 13-4**. The radials wires ideally would be uninsulated to make maximum contact with the Earth's surface. Each radial is typically $\lambda/4$ to $\lambda/2$ in length. Over typically lossy

ground, the antenna's effectiveness is a direct function of radial quantity. The resistance of R_G as measured by Jerry Sevick, W2FMI, in typical New Jersey soil is shown in **Fig 13-5**.

Note that the ground resistance approaches 0 Ω and thus the total load resistance approaches the value of R_R as the quantity reaches 100 radials. This requires a large quantity of wire, not to mention a significant amount of effort with a digging tool; however, that is exactly what is done by most commercial AM broadcast stations, which are typical monopole users. In addition, broadcasters tend to search out areas with high conductivity, so when you see such an antenna in a swamp, it isn't just because the real estate cost was low!

While it might be convenient to

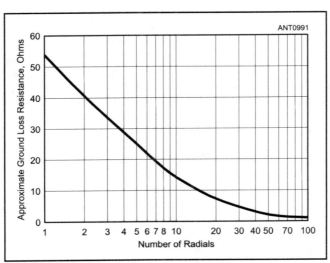

Fig 13-5 — Measured resistance of resonant monopole of Fig 13-4 as a function of the number of buried radials over typical soil.

use a small number of radials, let's say three, a look at Fig 13-5 indicates that the antenna impedance in his soil would be about 72 Ω. If it were, in fact, 72 Ω, R_G would be: R – R_R = (72 – 36) or 36 Ω. The resulting efficiency, per Eq 13-3, is thus 36/72 = 50%, corresponding to a signal loss of 3 dB, or half your power. Note that a nice match to 50 Ω cable occurs with 10 radials in this soil. While that is convenient, it also implies that R_G = 14 Ω. The resulting efficiency is thus 36/50 = 72%, or a signal loss of 1.4 dB.

For any desired efficiency, it is conceptually a simple matter to add ground radials until R_G is reduced to a value that results in an acceptable efficiency. There is often an easy trade-off between the cost of increasing transmitter power and the cost of additional radials. While the efficiency impacts both transmit and receive performance, in the lower HF region, the receiver performance is generally limited by received noise, which is reduced along with the desired signal, making the improvement moot.

There are some services, for example Amateur Radio, in which government regulations limit transmitter output power. In that case, once the limit is reached, improving antenna efficiency may be the only

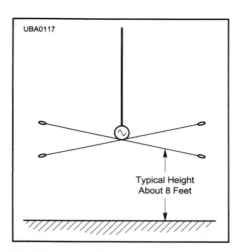

Fig 13-6 — Configuration of a monopole with an elevated ground system.

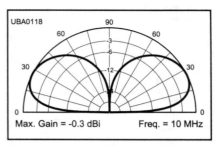

Max. Gain = -0.3 dBi   Freq. = 10 MHz

Fig 13-7 — Elevation pattern of λ/4 monopole over typical ground.

radials. Unfortunately, some of the benefits of the monopole are lost in the process, including its low visibility and lack of required dedicated real estate. Considered from another point of view, a monopole with four elevated radials requires exactly the same hardware and number of supports as two dipoles, not counting the monopole itself. If the supports can be made somewhat higher, in most cases (except perhaps on the oceanfront) the resulting dipole system will be more effective.

The Ground as a Reflector and Attenuator for the Far Field

A somewhat more subtle, but often much more significant, ground effect occurs particularly with respect to the low-angle reflection leaving the monopole. We often select a monopole because we would like long-distance propagation by low angle radiation; however, ground conductivity impacts us in two ways.

Lossy ground actually slows down the radiation in contact with the Earth's surface, resulting in a tilt to the wavefront that makes it stay in closer contact with the ground for greater distances before it is launched into space. This effect can be of benefit to medium-wave AM broadcasters, who obtain extended ground wave coverage as a result.

For those interested in the wavefront leaving the surface and launching skyward, the extended proximity to the lossy ground results in signals being attenuated by heating of the ground. This is compounded by the fact that lossy ground is a less effective reflector than a perfect ground or an ocean, so that the reinforcement at the horizon that we would like from the reflected wave is less effective than it might be.

Unfortunately, once we pick an antenna location, the ground at some distance from our antenna is even less under our control than that underneath the antenna. The result of lossy ground in the vicinity of the antenna can be seen in **Fig 13-7**, an *EZNEC* predic-

tion of a monopole with eight radials over typical ground. Note that, unlike the perfect-ground case, shown in Fig 13-3, there is no resultant radiation at the horizon and all the low-angle radiation is reduced.

You can construct an extended ground system extending many wavelengths from the antenna base to support the very low angle radiation; however, the amount of real estate required is quite large. I have seen this done on one occasion and it worked well; however, it made use of an abandoned Air Force base, and had (surprise) a large budget.

It is interesting to compare this typical monopole with a dipole at the same height as the top of the monopole (λ/4). The two patterns are shown together in **Fig 13-8**, considering the broadside response of the dipole — it isn't quite omnidirectional, even at this height. The dipole has about the same signal at the low angles, in its best direction, as well as considerably more at the high angles. The reason for this comparison is that you clearly can get at least one support that high! Of course, a higher dipole would have significantly more low angle response.

choice in order to achieve optimized transmit effectiveness from a monopole. There may be other issues to deal with. I once worked on an offshore system with vertical monopoles in which some local nationals would steal the radials at night to sell for scrap. There, we had a distinct incentive to find a way to optimize using fewer radials!

Achieving High Efficiency with an Elevated Monopole

A typical ground system for feeding power to a vertical monopole suffers from the losses due to the conductivity of imperfect soil. One way to improve on this is to elevate the monopole above ground and create an artificial ground structure with minimal losses. This can be accomplished by having multiple resonant insulated radials, each tuned to be λ/4 long, at some height above real ground.

Usual practice is to have the radials at least 8 feet above ground to avoid human strangulation or deer-antler interference (some areas may require greater heights or perhaps fencing). The configuration is shown in **Fig 13-6**. If raised well above the Earth, this antenna is called a *ground plane* and is quite popular as a VHF omnidirectional fixed-station antenna.

A monopole with an elevated ground system approaches the efficiency of a ground-mounted monopole with a large number of buried

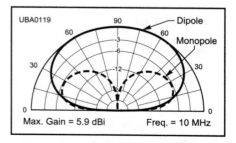

Max. Gain = 5.9 dBi   Freq. = 10 MHz

Fig 13-8 — Elevation pattern of λ/4 monopole over typical ground compared to horizontal dipole λ/4 above typical ground.

The monopole has a null directly overhead — which is the dipole's maximum elevation angle — sometimes an advantage for either antenna, depending on your objective. In some cases, received noise comes from medium range sources that propagate at high angles. This often provides a significant advantage in received signal-to-noise ratio if the desired signal arrives at a low angle. In this application, the monopole shines, although there are even more capable antennas that will be covered later in the book.

Wrap Up

A ground-mounted vertical monopole may be a logical choice for an HF antenna, especially if omnidirectional coverage is desired. It often can be less obvious aesthetically than a horizontal antenna, and perhaps even be disguised as a flagpole or a rain gutter. A vertical also takes up very little real estate — once the ground system is installed, of course. It also can really shine in performance — especially if its base is right at the water's edge.

Monopoles are quite useful as elements in multielement HF arrays, as will be discussed in the next chapter. They offer the possibility of effective electronic beam steering in azimuth — often difficult to do with other types of elements.

Review Questions

13-1. Under what conditions can a ground mounted monopole be more effective than a horizontal antenna?

13-2. What are the major disadvantages of a monopole?

13-3. What problem does a monopole with an elevated ground solve? Why might it not be a good trade-off?

Chapter 14

Arrays of Vertical Monopole Antennas

The fields from multiple monopoles can be combined to add in desired directions. This array uses four vertical elements on corners of a square.

Contents

Arrays of Vertical Monopole Antennas

One of the benefits of a vertical monopole discussed in the previous chapter is its omnidirectional azimuth radiation pattern. This is ideal in some applications, such as broadcasting or for a base station that needs to communicate with mobile stations in many directions. The other side of that coin is that the radiation intensity in any particular direction is less than would result from most horizontal antennas with their limited azimuth coverage. However, by using multiple monopoles arranged in arrays virtually any desired azimuth coverage can be obtained with resulting gain in desired directions. In some cases the same physical antenna array can be aimed in different directions in response to changes in requirements.

Lining Up Elements of a Vertical Array

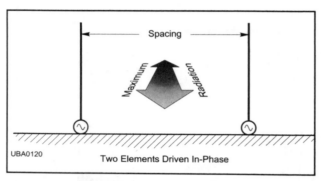

Fig 14-1 — Configuration of a broadside array of two monopole antennas.

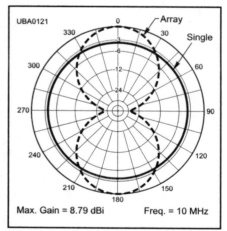

Fig 14-2 — Azimuth pattern of a broadside array of two monopole antennas with λ/2 spacing (dashed line) compared to that of a single monopole (solid line).

Max. Gain = 8.79 dBi   Freq. = 10 MHz

The directivity of an array of monopoles will be determined by the quantity, spacings and phase differences of the applied signals. I will discuss a number of different cases, starting with perhaps the most straightforward.

Two Element Broadside Array

This two-element broadside array consists of two monopoles, fed by equal in-phase signals. **Fig 14-1** shows the configuration. The signals combine at maximum strength along a line equidistant from both monopoles, which is perpendicular to the center of a line drawn between the elements.

The azimuth radiation pattern of such an array with a spacing of λ/2 is shown in **Fig 14-2**, compared to that of a single monopole. These patterns are for the ideal case, over perfectly conducting ground. All the effects of ground conditions relating to efficiency and far-field ground losses discussed in the previous chapter also affect the radiation from the array, to the same extent as the single monopole. The elevation pattern is essentially unchanged from the single monopole.

Note that the spacing of λ/2 is a special case in that the radiation from each element, while starting in-phase with that of its neighbor, is 180° out-of-phase by the time it reaches the other element. This results in a deep null along the line between the elements.

At spacings below λ/2, the null fills in and the reduction in sideways rejection is much less, as shown in the λ/4 case in **Fig 14-3**. For spacings greater than λ/2, the side radiation actually increases — to a

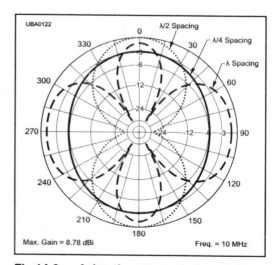

Max. Gain = 8.78 dBi   Freq. = 10 MHz

Fig 14-3 — Azimuth pattern of a broadside array of two monopole antennas with λ/2 spacing (dotted line) compared to λ/4 (solid line) and λ (dashed line).

maximum at a full λ, since the energy arriving at the other elements is now 360° out-of-phase, which is the same as being in-phase. Above λ/2 spacing, arrays have significant lobes at other azimuths, and thus most systems use spacings less than λ/2. One exception is a spacing of 5λ/8, which offers the maximum broadside gain at a cost of some spurious lobes. Its azimuth pattern is shown in **Fig 14-4**.

Two Element End-Fire Array

Just as in the case of horizontal dipoles, two monopoles fed equal currents 180° out of phase as shown in **Fig 14-5**, will have no radiation in the broadside direction. For spacing less than λ, maximum gain is in the plane of the elements. The gain increases for close spacings. The coupled mutual impedance subtracts and thus reduces the impedance at each element. **Fig 14-6** shows the pattern for spacings of λ/2 and λ/8. Any additional gain is small.

In most cases, spacings of less than about λ/8 start to show losses due to the low impedances that may offset the additional gain due to directivity. **Fig 14-7** shows the impedance of both broadside and end-fire arrays as a function of element spacing.

Two Element Cardioid Array

An interesting and useful array is one in which the elements are spaced λ/4 apart and fed 90° out-of-phase.

By the time the signal from the delayed element reaches the element with the leading phase, its phase is now 90° later and the two signals are in-phase and add. The signal from the leading element, on the other hand, reaches the lagging element with a delay such that the lagging element is now 180° behind its phase and thus they cancel in that direction. The result is a very useful unidirectional pattern with a deep null to the rear, as shown in **Fig 14-8**. A transmission line section with an electrical length of λ/4 can be used to yield the desired 90° delay. In some configurations the feed line can be remotely switched between the elements to provide reversible directionality.

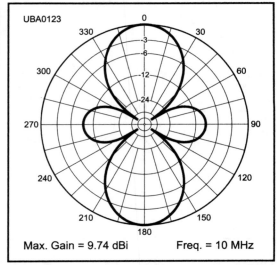

Fig 14-4 — Azimuth pattern of a broadside array of two monopole antennas with 5λ/8 spacing.

Max. Gain = 9.74 dBi   Freq. = 10 MHz

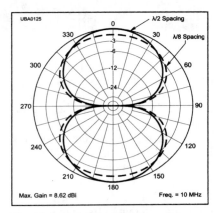

Max. Gain = 8.62 dBi   Freq. = 10 MHz

Fig 14-6 — Azimuth pattern of an end-fire array of two monopole antennas with λ/2 spacing (solid) compared to λ/8 (dotted).

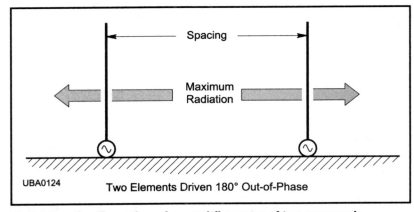

Fig 14-5 — Configuration of an end-fire array of two monopole antennas.

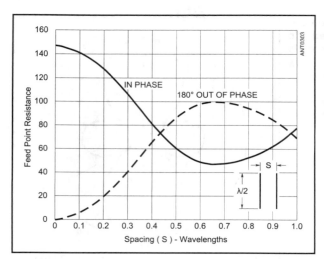

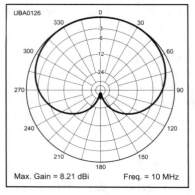

Fig 14-7 — Impedance of each monopole in broadside and end-fire configurations as a function of element spacing.

Fig 14-8 — Azimuth pattern of a cardioid array of two monopole antennas with λ/4 spacing and 90° phase shift.

Feeding Driven Arrays

There is an issue that should be raised at this point about feeding complex driven arrays. Any that are not fed either in-phase or 180° out-of-phase require special attention. The patterns shown assume equal currents in each element at the desired phase.

We have mentioned that elements have mutual coupling between them that change the driving-point impedance to that of the antenna element itself plus or minus the coupled impedance. This means that unless the resulting impedance of each of the elements is the same, there will not be the same match to the transmission line and thus unequal currents as well as the possibility of unanticipated phase shifts, if the element impedances have reactive components.

There are a number of ways to deal with this issue, many beyond the scope of this book, but a few should provide examples of some current practice.

• Many AM MF broadcast stations use multielement phased antenna arrays to reach selected regions of customers or to avoid interference to other users of the same channel. They generally have adjustable matching networks at the base of each element to allow precise adjustment of current amplitude

and phase. They often have remote measurement systems to determine the field strength in different directions to ensure that the adjustments are correctly made.

• A property of odd multiples of λ/4 lengths of transmission lines is that, if fed in parallel, they will deliver the identical currents to their loads — even if the load impedances for each line are different. This property can be used in some arrays to *force* identical currents to multiple elements. (See Chapter 8 in late editions of *The ARRL Antenna Book* for details.)

Steering the Beam From an Array of Monopoles

So far I have discussed all the extreme cases of feeding two driven antenna elements. In addition, intermediate phasing of elements can be used to steer a beam in virtually any direction. By delaying the phase of the signal to one element, it is (almost) equivalent to physically moving that element back (or the other one forward) by a distance corresponding to the signal delay.

Consider a signal delayed some fraction of a wavelength, in this ex-

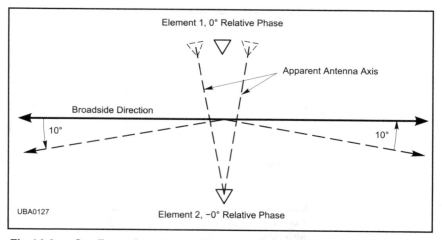

Fig 14-9 — Configuration of a broadside array with delayed phase in one element to steer beam by about 10°.

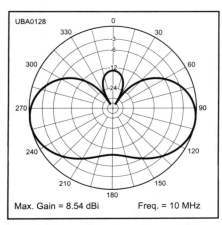

Fig 14-10 — Azimuth pattern of the steered array of Fig 14-9.

ample λ/12 or 30°. At a wavelength of 30 meters, that corresponds to a distance of 30/12 or 2.5 meters. Our elements are spaced λ/2 or 15 meters apart. This is equivalent to shifting the axis of the antenna by an angle of $\sin^{-1}$ (2.5/15) or 9.6°. See **Fig 14-9**. The resulting pattern is shown in **Fig 14-10**.

In many applications, it is desired to have a steered unidirectional pattern. The steered array of Fig 14-9 can have additional elements λ/4 behind them, phased by 90° from each of the originals, to result in a pair of cardioid arrays with a beam steered as shown in **Fig 14-11**.

Arrays With Unequal Currents

So far, all of the arrays examined have been fed with equal currents. This is generally the case with pairs of elements, or pairs of pairs, as in the steered-cardioid array in Fig 14-11. In the case of larger arrays, the width of the array often exceeds λ/2, resulting in undesired directions at which signals from the further-apart elements add in-phase. Consider an array consisting of eight vertical monopoles each separated by λ/4 with the intent of feeding them in-phase to obtain a narrow high-gain beam broadside to the array. The azimuth pattern of such a system is shown in **Fig 14-12**.

Note that the design objectives have been met. The main beam is less than 26° wide on each side and the

gain is 12.5 dBi, a very respectable signal. Notice something else, the spurious *sidelobes*. These have cropped up because of the in-phase combinations of the outer elements at 45° off axis on each side. For many applications these extra lobes are a small price to pay for the additional gain of the larger array. There are some applications, however, in which such types of sidelobes are completely unacceptable. For example, in a radar system with a rotating antenna, a single aircraft would seem to appear at different azimuths (with different strengths), resulting in false alarms and mistargeting of weapons. A modern digital system with processing can be used to eliminate the false targets; however, it is also possible to eliminate them through a change to the antenna design based on adjusting the currents in each element.

The resulting system is called a *tapered* current distribution. The idea is that the currents delivered to the outside elements are reduced in a systematic manner to deemphasize their contributions to the sidelobes, while still providing some contribution to the desired main beam. There are a number of systems in use; I will briefly describe two:

Linear taper — the currents in each element are reduced linearly starting in the center and tapering down towards each end. The slope of the taper can be adjusted with a steep slope having the most benefit, but also reducing the gain of the main beam the most. **Fig 14-13** shows the effect of applying a taper of 1, 2, 3, 4, 4, 3, 2, 1 across the array. Note that the sidelobes are almost, but not quite, gone. The beamwidth has increased by about 7° and the gain is down about 1 dB.

Binomial taper — The use of binomial coefficients as a tapering function can be shown to virtually eliminate sidelobes, albeit at a greater cost in desired characteristics. The binomial coefficients for an 8-element array are 1, 7, 21, 35, 35, 21, 7, 1. The azimuth pattern of

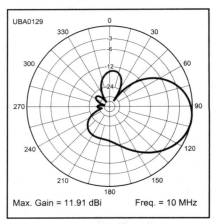

Fig 14-11 — Azimuth pattern of the steered array of Fig 14-9 with added cardioid shaping elements.

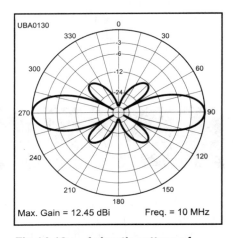

Fig 14-12 — Azimuth pattern of eight element broadside array with uniform current distribution.

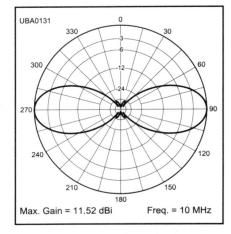

Fig 14-13 — Azimuth pattern of eight element broadside array with linear tapered current distribution.

such an array with that current distribution is shown in **Fig 14-14**. Note that there are very low sidelobes; however, the deep side nulls are gone and the beamwidth is significantly wider, reducing the gain.

Binomial Coefficients

Binomial coefficients are frequently encountered in antenna (as well as filtering) systems. They are the constants preceding each variable term found by performing the algebraic raising of a binomial to a power as in $(a + b)^n$. For example, if n = 2, $(a + b)^2 = 1 \times a^2 + 2 \times a \times b + 1 \times b^2$, and the coefficients are 1, 2, 1. If n = 3, $(a + b)^2 = 1 \times a^3 + 3 \times a^2 \times b + 3 \times a \times b^2 + 1 \times b^3$, and the coefficients are 1, 3, 3, 1.

While the coefficients can be determined by algebraic multiplication, a shortcut is to observe that if they are put above each other starting with n=1, any coefficient can be found by adding the term above to the one to the left of the term above as in the following:

1, 1
1, 2, 1
1, 3, 3, 1
1, 4, 6, 4, 1
1, 5, 10, 10, 5, 1
1, 6, 15, 20, 15, 6, 1
1, 7, 21, 35, 35, 21, 7, 1
and so on.

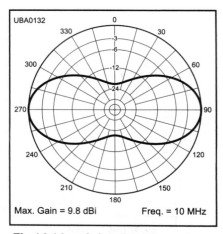

Fig 14-14 — Azimuth pattern of eight element broadside array with binomial tapered current distribution.

Chapter Summary

This chapter has introduced some interesting and significant concepts in the capabilities of antennas. We have seen that multiple vertical monopoles can have their patterns shifted in various ways — broadside, end-fire, cardioid and even towards almost any azimuth by shifting the relative phase of the currents in each element. This allows the possibility of electronically steered antenna arrays, a very powerful tool in the toolbox of the antenna designer.

Review Questions

14-1. If you feed two monopoles with equal lengths of coax cable, which way will the pattern point?

14-2. If you add a $\lambda/2$ section of transmission line to one of the monopoles in question 14-1, what change will occur to the pattern?

14-3. If you feed two monopoles spaced $\lambda/4$ with equal currents 90 out-of-phase, what happens to the resulting pattern?

14-4. Why might you want to feed a large array with non-uniform element currents? What are the trade-offs involved?

Chapter 15

Practical Multielement Driven Arrays

These multiple elements are combined to achieve an omnidirectional pattern.

Contents

Practical Multielement Driven Arrays

In earlier chapters I have discussed a number of arrangements of horizontal and vertical elements used as single and multi-element antenna arrays. Almost any combination of them, in almost any quantity, can be (and probably has been) used to construct antennas that meet specific requirements. In this chapter I will discuss how principles described in preceding chapters have been effectively used to construct some classic antenna arrays.

The Bobtail Curtain Array

One of the simplest, and perhaps oldest, three element vertical monopole arrays is called a *Bobtail Curtain*. It consists of three vertical monopoles fed in-phase to form a broadside array with a tapered current distribution. The construction shown in **Fig 15-1** is deceptively simple, but it has some unusual features in spite of its simplicity.

The three monopoles are fed via a single connection to the bottom of the center element. Note that the outer two verticals are not connected to ground. This makes the bottoms the high voltage end of each outer monopole. The center vertical also exhibits a high impedance at the bottom, by virtue of its being connected to the other two through the lengths specified. It thus requires a transformation to match to low-impedance transmission line. In its earliest days, this transformation was generally accomplished with a link-coupled tuned circuit. The link would form the low-impedance driving point, while the resonant circuit would be resonated at the desired frequency. The antenna was connected to one end of the parallel-tuned resonant circuit, and there was a connection to a ground system on the other end. These days, a weatherproof remote automatic antenna-tuning unit is commonly used for matching.

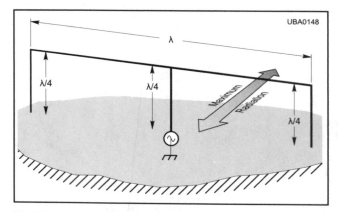

Fig 15-1 — Configuration of a Bobtail Curtain three element monopole broadside antenna system.

The high-impedance feed results in the maximum current being at the top of each monopole. This raises the area of maximum radiation away from ground and makes the efficiency somewhat less dependent on the conditions of the ground beneath the array. The effect of ground conditions beyond the near field is still the limiting factor for far-field radiation in the far field, as with any vertical HF array. Note that having the high-voltage end near the ground suggests that proper protection should be employed to avoid accidental contact by humans or animals during operation.

The low impedance point at the top of each outer vertical is connected in parallel to the center one through a $\lambda/2$ section of wire, so the impedance is low again at the connection point. Because all the current flows up the center element and then splits to head towards the outer elements,

each outer element has half the current of the center. Thus we have the kind of taper described in the last chapter. Note that with three elements, the taper will be $1 - 2 - 1$, which can be considered either a linear or a binomial taper, if you worry about naming such things!

The horizontal wires would also radiate, but because they are carrying identical currents in opposite directions, their radiation broadside to the elements cancels, while the nulls from the ends largely takes care of the rest.

Because there is a low impedance at the top, it is also possible to feed the junction at the top of the center element directly with a low-impedance transmission line. This is not usually done, however, because it is often difficult to install a transmission line without coupling it to the other portions of the antenna. The modeled impedance at that point is a close match to the usual 50 Ω coaxial cable, although there really isn't quite a ground at that point, so an isolating choke or balun would have to be employed to keep common-mode current off the coax shield.

How Does it Work?

The Bobtail Curtain is quite directive, as shown in **Fig 15-2**, with significant gain compared to a single monopole. The sharpness of its azimuth pattern makes it well suited for point-to-point links. Or you could use a number of Bobtail Curtains for wider coverage — especially if multiple supports are available.

The large horizontal section might make a reader wonder about the relative benefits of this antenna compared to a full-wave center-fed horizontal antenna hanging from the same supports. The horizontal antenna has almost the same azimuth pattern, as shown in **Fig 15-3**. However the choice between the two is clearly highlighted in **Fig 15-4**, which compares their elevation patterns over typical ground. Note that the horizontal antenna actually has a higher peak gain; however, it is generally directed at higher elevation angles. Note that the gain of the vertical curtain is significantly higher at lower elevation angles. The choices are thus:

• If you are designing an HF system for medium-range communication (typically 1000 miles or less) by means of *Near Vertical Incidence Skywave* (NVIS) propagation, the horizontal system would be more effective.

• On the other hand, if you want longer-haul communication at low takeoff angles, the vertical array will both provide stronger signals near the horizon and reduced interference and noise from closer-in stations and noise sources, often a limiting factor in reception.

For many installations seeking long-haul propagation, the issue often becomes height of available, or feasible, supports. At the lower frequencies in the HF or MF region, even λ/4 heights can be challenging, and a horizontal antenna capable of producing low-angle radiation requires heights well above λ/2. For example, at 1.8 MHz a λ/2 high antenna would be 273 feet high!

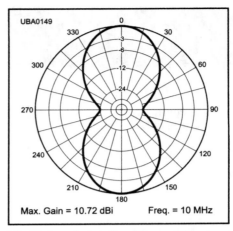

Fig 15-2 — Azimuth pattern of a Bobtail Curtain array at 10° elevation angle over typical flat ground.

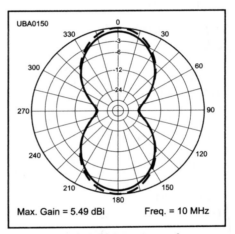

Fig 15-3 — Azimuth pattern of a Bobtail Curtain (solid line), compared to a full-wave center-fed horizontal antenna (dashed line) over typical flat ground at 30° elevation.

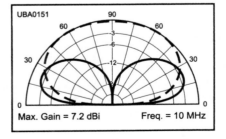

Fig 15-4 — Elevation pattern of a Bobtail Curtain (solid line), compared to a full-wave center-fed horizontal antenna (dashed line).

The X-Array Antenna

Another simple array to construct is the combination collinear-broadside system known as the *X-Array*. This antenna, shown in **Fig 15-5** is made from four folded dipoles fed in-phase. The dimensions between the elements are not terribly critical; however, those shown result in a good balance between gain and a clean pattern. They also result in a combined drive-point impedance of close to 50 Ω, a good match to coax, if each element feed line is an electrical full-wave and the parallel combination is fed through a balun to transition from balanced-to-unbalanced feed lines.

Building an X-Array Antenna

The only critical part of construction of this antenna is to make sure that all elements are tied together in-phase. This means the all right-hand-side element connections need to be on one side of the combined feed, while all left-side connections be on the other. Note that because the two wires in each folded dipole are at the same phase, it doesn't matter if the connections are to the top or bottom wire of the folded dipole elements.

By having the individual element feed lines (typically 300 Ω line is used) be exactly one electrical wavelength long, the folded-dipole feed impedance is repeated at the junction point. The effects of mutual coupling reduce the folded-dipole impedances so that the combination is very near 50 Ω.

You can make construction and materials provisioning easier if both antenna elements and dipole feed lines are all made from 300 Ω TV type twinlead or amateur type window line. You should limit the use of TV twinlead to low-power transmitters.

X-Array Performance

Azimuth and elevation patterns for an X-Array with elements at λ/2 and a full wavelength above ground

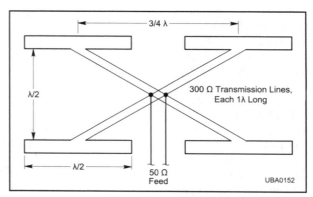

Fig 15-5 — Configuration of an X-Array of four folded-dipole elements.

are shown in **Figs 15-6** and **15-7** respectively. Note that the benefit of the ground reflection adds significantly to overall gain, as compared to vertical arrays, such as the Bobtail Curtain.

While the Bobtail Curtain is often employed at the lower end of the HF spectrum, the X-Array is an antenna best suited for the upper portion of the HF region into VHF because it requires significant vertical spacing. The antenna modeled for the following plots had the lower element at a height of λ/2 and the upper element another λ/2 high. At a 10 MHz modeling frequency, this represents a height of about 100 feet. While that height can be accommodated, the expense of two towers of that height argues for antennas that can be supported by a single support, such as I will cover in later chapters.

At 30 MHz, the top of the HF range, a height of 30 feet will do the trick and is not nearly as daunting. Existing trees or buildings can often be used to good advantage. As with the Bobtail Curtain, the X-Array has a narrow azimuthal pattern, best used for point-to point or in combination with other antennas. Many amateur stations using long, single-wire antennas have deep nulls in their coverage at some

azimuth angles, particularly at the higher frequencies. If the nulls are in an unfortunate direction, an antenna such as the X-Array can be an effective and simple way to fill in the gap.

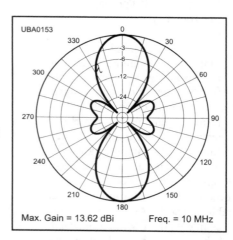

Fig 15-6 — Azimuth pattern of an X-Array at 17° elevation angle.

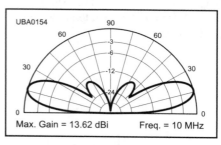

Fig 15-7 — Elevation pattern of an X-Array.

The Four-Square Array

An antenna that is very popular with amateur operators looking for long-haul performance on the 160 through 40 meter amateur bands is called the *Four-Square* array. This is a unidirectional array consisting of four, usually λ/4 long, monopoles at the corners of a square with sides λ/4 long. With the phasing of the sources shown in **Fig 15-8**, the antenna will fire on the diagonal between the elements with 0° and –180° phase delay.

The resulting azimuth and elevation patterns over typical earth are shown in **Figs 15-9** and **15-10**. Note that the half-power beamwidth of the azimuth pattern is 100°, so that four directions can cover all azimuths with some overlap. The real benefit of the Four-Square comes from its symmetry. Because the element spacings are the same, the feed system can be moved instead of the antenna elements and thus, via a relay network, all four directions can be achieved with a single antenna. The gain over an isotropic radiator is about 5.2 dBi over typical ground, although the gain would increase to 10.8 dBi if the ground were perfect. As with all vertical systems, ground conductivity some wavelengths from the system plays a large part in determining effectiveness at low takeoff angles.

It is interesting to compare the Four Square to the two element cardioid discussed in the last chapter. That is sort of half a Four-Square, with a half-power beamwidth of 170° and a peak gain 2.6 dB less than the Four-Square. It can cover almost all azimuths from one direction or the other, and adds the prospect of a less-than-optimum broadside array to fill in the gap. The switching is much easier to implement, however, without implementing the broadside direction.

Feeding the Four-Square

As noted in the last chapter, the ef-fects of mutual coupling on elements fed at different phase angles generally results in different feed impedances at each element. In order to force equal currents into each element at the desired phase, a combination of selected transmission-line lengths and an LC impedance-matching network can be employed. **Fig 15-11** shows the arrangement used to obtain the desired results from a Four-Square.

The exact values for the inductive and capacitive reactances shown in Fig 15-11 will depend on the ground conditions, the number of radials used at the base of each monopole and the Z_0 of the cable used for the phasing lines. Note that the electrical, not physical length of the cables is specified. These should be trimmed for the center of the desired band using an antenna analyzer or impedance bridge for optimum performance. The values for various arrangements are given in **Table 15-1**. These should be considered starting points, since the exact ground conditions will have a major impact on the impedance.

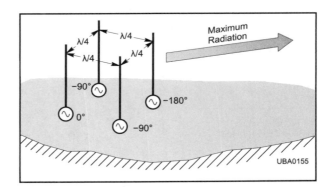

Fig 15-8 — Configuration of a Four-Square array of four λ/4 monopole elements, spaced λ/4 on each side.

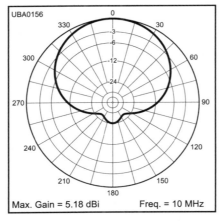

Max. Gain = 5.18 dBi   Freq. = 10 MHz

Fig 15-9 — Azimuth pattern of a Four-Square array at 25° elevation angle.

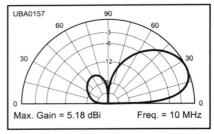

Max. Gain = 5.18 dBi   Freq. = 10 MHz

Fig 15-10 — Elevation pattern of a Four-Square array.

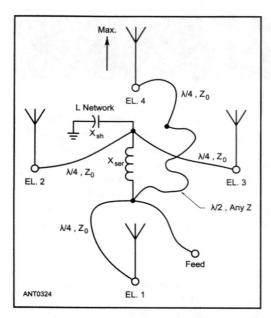

Fig 15-11 — Optimum feed arrangement for Four-Square array.

Table 15-1

L Network Values for the Four-Square Array Feed Arrangement of Fig 15-11.

Radials per Element	Cable Z_0	X_{SER} (Ω)	X_{SH} (Ω)
4	50	17.1	−13.7
4	75	38.5	−30.9
8	50	20.2	−15.6
8	75	45.4	−35.2
16	50	23.6	−17.6
16	75	53.1	−39.6

A Real-World Steerable Unidirectional Array

The eight element broadside array discussed briefly in the last chapter was part of a larger system I worked on early in my engineering career. By pairing each element with a rearward element phased to generate a cardioid response the system becomes a 16 element unidirectional broadside array with the azimuth pattern shown in **Fig 15-12**.

While not likely a system for the home constructer, this antenna was something that was successfully deployed in the field and can serve as an interesting example of what can be accomplished. This antenna was part of a world-wide system with separate receiving and transmitting systems located continents apart. There were four transmitting stations and five receiving locations. Each receiving station had to monitor all four transmitters, each at different azimuths.

The four transmitters were in the same quadrant, so rather than erect four separate antennas at each receiver location, a single array was used to synthesize four pencil-shaped beams — one

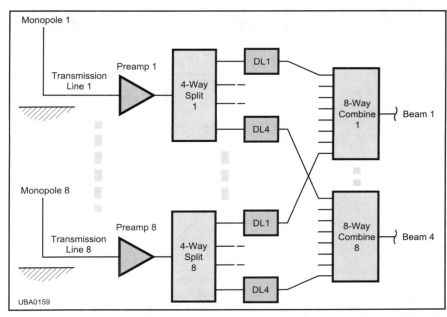

Fig 15-13 — **Configuration of an eight element broadside array that provides four simultaneous beams.**

towards each transmitter. This was accomplished by making the array unidirectional (using a reflecting screen to be discussed in a later chapter) to eliminate the unneeded response from the rear. The cardioid arrangement used here will provide the same result for a single frequency.

Each receiving element was fed through an equal-length transmission line to a preamplifier. Each preamplifier had a splitter with four outputs. Each output had a delay line corresponding to the shift for a single beam, as discussed for a two element array in the last chapter. The eight signals for each of the beams were combined in an eight-port summing network. Thus each beam was available for the appropriate receivers. The configuration is shown in **Fig 15-13**. **Fig 15-14** shows the azimuth response for three such beams, each 10° apart.

I was not only involved in the development of procedures to make sure that all phases were correct, but

had an opportunity to participate in an evaluation of the results at one location to validate the concept. A helicopter was fitted with a suspended vertical dipole with a battery-powered transmitter at its center. An observer with a transit was stationed at the

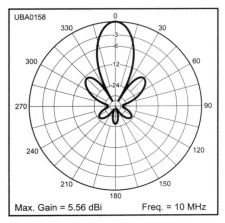

Fig 15-12 — **Azimuth pattern of a 16 element unidirectional broadside array showing three separate beams at 10° increments.**

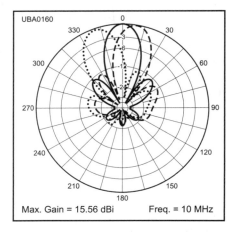

Fig 15-14 — **Azimuth pattern of the steered array in Fig 15-12, showing three separate beams at 10° increments.**

array center to call off azimuth positions while the helicopter's navigator used a hyperbolic navigation system to verify the radial distance. The signal from each receiver was measured and the antenna worked as specified (much to my relief).

Long-Wire Antennas

One of the simplest antennas to build is called a *long wire*. As its name implies, it is just a long, in terms of wavelengths, piece of wire. It can be resonant or not, although is often easier to feed and predict the performance if it is resonant. While some call any piece of wire a long wire, most would say it would have to be at least 2 λ long to qualify for the term.

The Basic Long Wire

A piece of wire 2 λ long is a kind of collinear array consisting of four λ/2 elements one after the other. Unlike the previously discussed multielement collinear antennas in which elements were fed in-phase, the end-to-end connection results in alternate sections being of opposite phase. Thus unlike the in-phase collinear with its main lobe broadside to the wire, an end-fed long wire with an even number of elements will have a broadside null and a pattern that has offset major lobes, as shown in **Fig 15-15**.

This wire can be fed at one end using high-impedance transmission line, or the first λ/2 element can be fed in the center with lower-impedance line. Its resonant impedance at that point is about 120 Ω, so a good match to

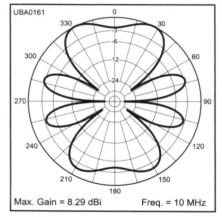

Fig 15-15 — Azimuth pattern of the 2 λ end-fed long wire.

50-Ω coax can be obtained through a λ/4 section of 75-Ω coax.

If the long wire is end fed (often connected directly to a tuning unit at the transmitter), it can provide good performance from a frequency as low as λ/4 (if fed against a good ground) and as high as desired. As the frequency is raised, the pattern becomes more complex and tends to focus closer to the line along the wire axis. For any given direction, there is likely a frequency that will provide good coverage.

It's hard to imagine an antenna with this amount of flexibility that is as easy to construct!

The V-Beam Array

If more directivity is desired than that afforded by a single long wire, you can configure two of them in the configuration shown in **Fig 15-16**. This is known as a *V-beam*. You take advantage of the fact that for a 2-λ wire, the center of the main lobe is 63° away from the broadside direction, or 27° from the wire axis. If you

combine the effects of two such wires fed in-phase with a separation of 54°, the main lobes will reinforce along the direction of the bisector of the angles, as shown in **Fig 15-17**.

The V-beam does provide usable bidirectional gain and directivity, combined with a nice balanced feed suitable for high-impedance low-loss balanced transmission lines. It can be operated on other frequencies, although the angle between the wires will only be optimum at its design point. For longer-length wires, the main lobes move closer to the wire axis and thus a smaller angle is beneficial.

The Rhombic

If you place two V-beams end-to-end, you would end up with the *rhombic* shown in **Fig 15-18**. This configuration further reinforces the on-axis directivity and further reduces the broad sidelobes of the V-beam, as shown in the azimuth pattern in **Fig 15-19**. The rhombic, however, really moves into prime time and

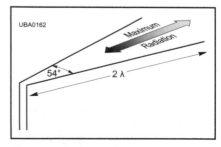

Fig 15-16 — Pair of 2 λ end-fed long wires combined into a V-beam.

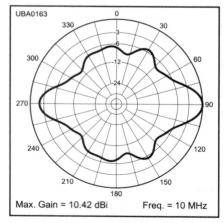

Fig 15-17 — Azimuth pattern of the 2 λ per leg V-beam.

changes character with the simple addition of a terminating resistor across the gap at the far end, as shown in **Fig 15-20**.

The *terminated rhombic* is a bit different in operation compared to antennas you've looked at. You might consider it as a special sort of transmission line. As previously mentioned, an open-wire transmission line doesn't radiate if the two wires are very close together. A ter-minated rhombic, however, you have essentially a terminated transmission line that has its wires spaced very far apart. The currents moving down the line to result in radiation, and the angles and spacings add in the desired direction along the bisector of the angle between the fed ends.

Note that, unlike other antennas you have studied, there is no reflection from the far end of the terminated rhombic. The result is that there is no radiation back towards the source end. You thus have a unidirectional antenna with significant gain and directivity, as shown in **Fig 15-21**. Terminated rhombics have long been used for fixed (for HF they are hard to rotate!) point-to-point transoceanic links and have provided good stable results. The terminating resistor absorbs about half the incident power and thus must be a substantial resistor for high-power transmitting applications.

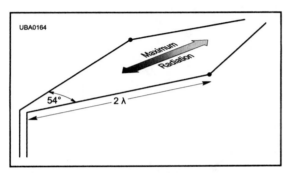

Fig 15-18 — A pair of two 2 λ V-beams combined into a rhombic.

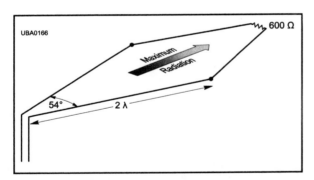

Fig 15-20 — A pair of 2 λ V-beams combined into a terminated rhombic.

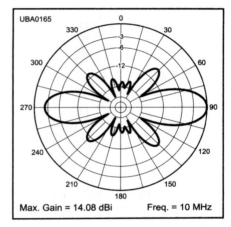

Fig 15-19 — Azimuth pattern of the 2 λ per leg rhombic.

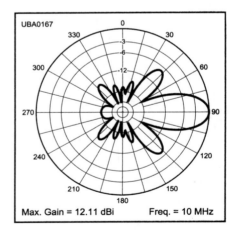

Fig 15-21 — Azimuth pattern of the 2 λ per leg terminated rhombic.

But Wait There's More!

Because of the lack of reflections from the termination point, the rhombic acts like a reasonably frequency independent load to the source and thus has a wide impedance bandwidth. The 600 Ω SWR over a 2:1 frequency range is shown in **Fig 15-22**. While the wire angles will not be optimum for multiple frequencies, the antenna can still be used over this frequency range with excellent results. At the high end, the pattern tends to break up; however, compromise angles can be found that can control this over a wide range. Just using the 10 MHz design angle used for the V beam, I achieved the results in **Fig 15-23** for 8 and 14 MHz operation.

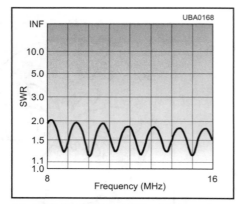

Fig 15-22 — 600-Ω SWR of 2 λ per leg rhombic over 2:1 frequency range.

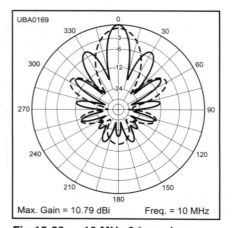

Fig 15-23 — 10 MHz 2 λ per leg rhombic's azimuth pattern at 8 MHz (dashed line) and 14 MHz (solid line).

The Beverage Long-Wire Receiving Antenna

An antenna that looks like a long wire and is made from a long piece of wire, but is quite different in operation from those we've discussed, is called a *Beverage* antenna. This is named for Harold Beverage, who developed and patented this antenna just after WW I. The Beverage antenna is designed to be a highly directive receiving antenna that responds to vertically polarized low-angle waves approaching along the ground. It accomplishes this by itself being installed quite close to the ground, cancelling most of its response to sky waves as well as horizontally polarized signals. Vertically polarized waves coming along the lossy ground tend to slow along the bottom near the ground, resulting in a tilt to the wavefront. This tilt couples a signal along the length of the wire that adds in-phase with the signal as it continues to propagate along the wire towards the matching transformer.

Waves coming from the other direction accumulate as well, but are dissipated in a terminating resistor, just like they do in the rhombic. The termination is usually between 400 and 600 Ω, adjusted for best front- to-back ratio. The configuration is shown in **Fig 15-24**.

The consequence of this is that desired signal is received without the noise from nearby lightning or other noise sources that tend to arrive from high elevation angles. Other interfering noise or signals that arrive from unwanted directions are also attenuated by the directional characteristics of the Beverage.

The absolute gain of a Beverage system is quite low; however, this receiving antenna is typically used for low frequencies (MF and just into HF) where atmospheric noise is the key limitation to signal-to-noise ratio and thus the weak desired signal is at much higher SNR than with other antennas because the noise is reduced more than the signal.

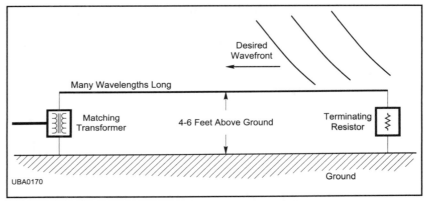

Fig 15-24 — Configuration of a Beverage low-noise receiving antenna.

Chapter Summary

This chapter has provided some easy — and some not-so-easy — examples of practical arrays that can be constructed to meet particular operating objectives. An interesting aspect of this is that most can be constructed from available materials at relatively low cost, compared to commercially available antennas, yet they all offer significant performance advantages over simpler systems.

Review Questions

15-1. Why might it be desirable to have the bottom of the Bobtail Curtain elements 8 feet above ground?

15-2. How many X-Array antennas would be needed to have coverage of all 360° at the –3 dB level?

15-3. How might you make a unidirectional version of an X-Array?

15-4. Why are extra components required to properly drive the elements of a Four-Square array at the proper phase and amplitude?

Chapter 16

Surface Reflector Antennas

This reflector antenna is used for point-to-point communication at 10 GHz.

Contents

Surface Reflector Antennas

As noted in previous chapters, it is frequently desirable to restrict the coverage of antennas to a single direction. I have previously introduced the λ/4 spaced, 90° phased array that produces a cardioid unidirectional pattern. There are other arrangements that have a similar effect.

Just as radio waves reflect from a large conductive surface, such as seawater, they will also reflect from a surface such as a metal plane. If the plane is large enough to reflect most of the radiation from the antenna, it will redirect the radiation away from the plane. If the spacing is such that the radiation ends up largely in-phase with the radiation from the antenna itself, this will reinforce the radiation from the antenna, resulting in a stronger signal in the desired direction and no signal to the rear.

The Plane Reflector Antenna

The simplest case to imagine is an infinite, flat and perfectly conducting plane. Fortunately it neither has to be infinite nor perfect to function quite nicely, thank you, but let's start here. **Fig 16-1** shows the configuration of a dipole placed in front of such a plane. You can imagine that this is a horizontal dipole above a perfectly conducting ground and it acts just the same way. Because the dipole is parallel to the surface, the phase of the reflected wave is out-of-phase with the incident wave, resulting in a phase shift of 180° at the boundary.

What Spacing Gives Us the Correct Phase?

You would generally like the signal coming directly from the antenna to reflect in-phase with the reflection so they add up going away from the reflector. If you space the dipole λ/4 in front of the reflector, the wave going rearward will have a propagation delay corresponding to a 90° phase shift as it travels to the reflector. The reflector will impart a 180° delay to the signal, as explained above, and there will be another 90° delay for the signal to return to the dipole, so that the total delay is 360°. This means that signals going directly rearward will add in-phase with signals leaving the dipole traveling towards the front.

Note that waves leaving the dipole at any other angles take a longer path and don't add fully in-phase. Nonetheless, the combined effect is beneficial. This is shown in **Fig 16-2**.

How Do You Make a Giant Plane at HF?

While a large solid metal plane is sometimes used at the higher VHF frequencies and above, at HF and for many applications above HF, a plate that extends outward on each side of the dipole by around λ/8 and above and below by λ/4 will have a similar effect. In addition, by replacing the plane with a *skeleton* of wires spaced much closer than a wavelength, the weight, cost and wind resistance can be reduced considerably.

In developing the models for these patterns, I wanted to get a feel for the effect of skeleton-wire spacing. Using my usual 10 MHz (30 meter) frequency, I started with a wire spacing of 1 foot, about 0.01 λ. I then reduced every other wire and continued until I reached a spacing of 4 feet, or about 0.04 λ, before I noted any degradation in forward gain or front-to-

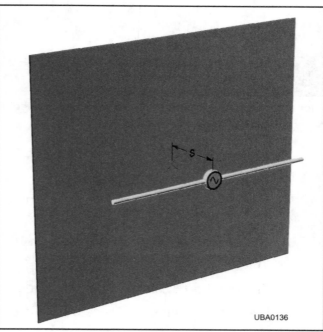

UBA0136

Fig 16-1 — Dipole in front of a plane reflector.

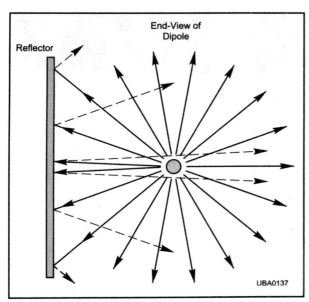

Fig 16-2 — Side view of dipole in front of a plane reflector. The direct radiation from the dipole is shown as a solid line, while reflected waves are shown dashed.

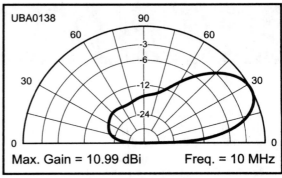

Fig 16-3 — Elevation pattern of the dipole in front of a skeleton plane reflector.

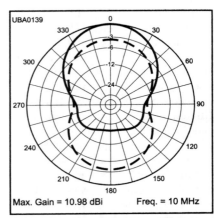

Fig 16-4 — Combined effect of all reflected waves from the plane reflector and the resulting pattern (solid), compared to a dipole in free space (dashed).

back ratio. This is probably a reasonable guideline for construction of reflector arrays, and is what I used for these models. Extending the width or height of the reflecting screen for the flat plane beyond these guidelines has very little impact on the predicted results.

How Does the Skeleton Play?

Fig 16-3 shows the elevation pattern of the skeleton-plane reflector over typical, real ground, with the center of the antenna at a height of $\lambda/2$. **Fig 16-4** shows the azimuth pattern compared to a dipole at the same height. Note that the radiation intensity off the front of the antenna is more than twice (about 3.6 dB) that of a dipole at the same height. Thus the rearward energy has been successfully redirected to refocus in the desired direction.

Had I used an infinite reflecting plane in free space (instead of a skeleton above earth), the forward gain compared to a dipole would have been an impressive 5.5 dB. The infinite plane would also have an infinite front-to-back ratio, compared to a respectable 20 dB for the skeleton. Unfortunately, infinite planes and free space are difficult to find, especially in the same place!

The Corner Reflector Antenna

A common alternative structure is the *corner reflector*. This can be created by folding the plane along the center behind the dipole so that the two sides are at an angle, as shown in **Fig 16-5**. The reflected waves tend to be focused somewhat more along the front axis, and with the reflected waves having a shorter distance to travel, they are closer to being in-phase with the front facing waves. The elevation pattern for a corner reflector with a 90° angle is shown in **Fig 16-6**, while the azimuth pattern, again compared to a dipole, is shown in **Fig 16-7**. For about the same size of plane (bent at the center), the corner reflector has a slight advantage over the plane reflector. However, the difference is small enough that mechanical considerations may be more of a consideration than performance when deciding between them.

Since the reflected waves travel different lengths, there is nothing magic about the λ/4 spacing, so you can use other spacings too. With narrower angles and longer plane sheets, you can achieve additional gain. A 45° corner with side lengths of around 2 λ should provide about 3 dB more forward gain, for example.

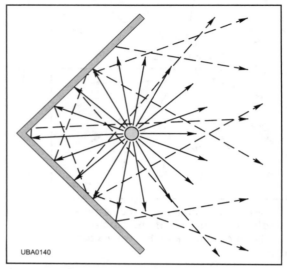

Fig 16-5 — Side view of dipole in front of a corner reflector. The direct radiation from the dipole is shown in solid, while the reflected waves are shown dashed.

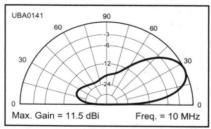

Fig 16-6 — Elevation pattern of the dipole in front of a skeleton corner reflector.

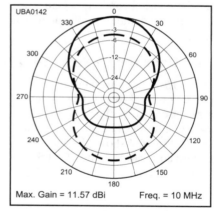

Fig 16-7 — Combined effect of all reflected waves from the corner reflector and the resulting pattern (solid), compared to a dipole in free space (dashed).

The Parabolic Reflector Antenna

Another structure you will sometimes encounter is the *parabolic reflector*. This is a special shape that has the property that all waves striking the reflector emanating from the *focus*, where you mount a dipole, leave the reflector in-phase and head out in parallel with the wave leaving the dipole in the desired forward direction. This is shown in **Fig 16-8**.

The power of this structure will be more evident later in this chapter when I discuss the three-dimensional paraboloid; however, in two dimensions, it is an alternate to the corner reflector. The elevation pattern for a parabolic reflector about the same size as a corner reflector is shown in **Fig 16-9**, while the azimuth pattern, compared to a dipole by itself, is shown in **Fig 16-10**.

Some Comments about Surface Reflectors

A reasonable question to ask about any of the above antennas is why bother? After all, a cardioid array can provide similar results with fewer wires. That is a good question and there are a number of answers to this, the relevance of which will depend on your design goals and constraints.

• The surface reflector is straightforward, predictable and easy to make work. Using this kind of structure avoids the problem of mutual impedance between multiple elements and possible complications obtaining equal currents in multiple elements.

• The same structure can be used at higher frequencies by just moving the dipole closer to the screen.

• It is easily adaptable to being driven in more complex structures, such as broadside arrays.

• The reflector, using the same screen (if a plane array), can operate in two directions simultaneously by having dipoles or driven arrays on each side.

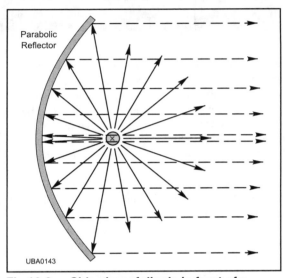

Fig 16-8 — Side view of dipole in front of a parabolic reflector. The direct radiation from the dipole is shown in solid, while the reflected waves are shown dashed.

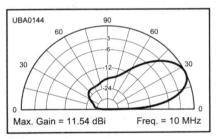

Fig 16-9 — Elevation pattern of the dipole in front of a skeleton parabolic reflector.

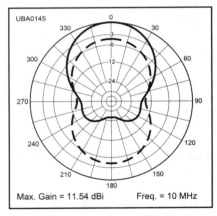

Fig 16-10 — Azimuth pattern of the parabolic reflector and the resulting pattern (solid), compared to a dipole in free space (dashed).

The Parabolic Dish Reflector — In Three Dimensions

So far, I have been talking about antenna structures in which the driving element is comparable in size to the reflector. As frequencies move into the upper UHF and especially microwave regions (above 1000 MHz), the wavelengths get small and antennas that would be out of the question at HF become feasible. Nowhere is this more evident than in the case of the *parabolic dish reflector*.

This is the same structure used as an optical reflector in most flashlights and vehicle headlights. Take a moment and observe a flashlight and note the difference in intensity in the *main lobe* of the flashlight compared to looking at the bulb from the side. This kind of dish shape can be just as effective for radio signals.

Because my usual 10 MHz dipole is relatively large, and because it is inherently omnidirectional around its axis, I've restricted the discussion so far to the two dimensional structure explored in the last section. While as effective as other two dimensional reflectors, the parabola really pays off when its driving antenna becomes small enough compared to the reflector size that it can be considered a point rather than a line. If that's the case, the parabola can be shaped like a dish and all radiation from the focus towards the dish will be sent from the reflector in-phase. This is perhaps more apparent in the case of a receiving antenna, in which virtually all the radiation received in the aperture of the disk is focused on the single focus point.

With a very small driving antenna, it becomes feasible to have a sub-reflector in front of the driving antenna to eliminate the direct radiation from the dipole or other driver. By eliminating the direct rays from the front of the driver, all the ra-diation can be via the parabola. Note that since you no longer have a direct ray to be in-phase with, there is no longer a need for a particular distance from the driver to the dish. You can use any dish size, with any focal distance, at any frequency with the appropriate driving antenna and sub-reflector located at the optical focus. You decide how much gain you need, or alternately how much receive aperture is needed for the desired signal strength, and then you select the appropriate-sized dish.

The down side of the sub-reflector is that it effectively provides a shadow to the signal to and from the main reflector. With higher frequencies, smaller drivers and larger dishes, the loss is generally a small fraction of the dish area. Typical gains for parabolic dishes are shown in **Table 16-1** for various frequencies and commonly available dish sizes.

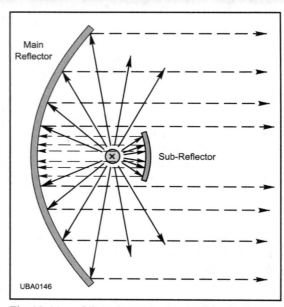

Fig 16-11 — Slice through a three-dimensional parabolic reflector antenna showing sub-reflector.

Complex Feeds are Common at Microwave Frequencies

While these discussions on reflector arrays have focused (no pun intended) on dipole feeds, at higher frequencies where we find most parabolic dish antennas, other types of drivers are more common. In the microwave region, perhaps the most common type is the *horn* antenna. This antenna, which will be covered

Table 16-1

Expected Gain of Parabolic Dish Antennas Compared to Isotropic Radiator (dBi).

Frequency (MHz)	Dish Diameter (Feet)						
	2	4	6	10	15	20	30
420	6.0	12.0	15.5	20.0	23.5	26.0	29.5
1215	15.0	21.0	24.5	29.0	32.5	35.0	38.0
2300	20.5	26.5	30.0	34.5	38.0	40.5	44.0
5600	28.5	34.5	38.0	42.5	46.0	48.5	52.0
10,000	33.5	39.5	43.0	47.5	51.0	53.5	57.0

later, is unidirectional by nature and can be designed to just radiate into the area of the dish without the need for a sub-reflector.

Dishes Get Quite Large

I have encountered rotating dish antennas with diameters larger than 100 feet. I have also seen sections of parabolic dish antennas as wide as 540 feet. These were used for a very-long-range space radar system, in which the drivers rotated to use different segments to cover different azimuths. There were two drivers at different elevation angles to simultaneously track targets crossing two elevation angles. This allowed computation of ballistic trajectories.

This system was reputed to be powerful enough (a combination of a massive transmitter and very high gain antenna) to detect a basketball-sized object thousands of miles away or to *shoot down* birds that got too close. The radar was located north of the artic circle where there weren't

many birds, but there were lots of warning signs posted for the occasional human that got near.

Perhaps the largest parabolic dish around is the one in use at the radio telescope of the Arecibo National Observatory in a remote corner of Puerto Rico. It has a fixed dish 305 meters (1000 feet) across that's built into a sinkhole in the rugged terrain. This is shown in **Fig 16-12**. This antenna is aimed by knowing what part of space this location is pointing toward as the earth turns and the cosmos moves around it. The large surface of this dish covers an area of more than 20 acres. It offers some 18 acres, about 26 football fields, of available receiving aperture! The dish reflects and concentrates weak celestial signals on the receiving antennas suspended 450 feet above its surface.

Fig 16-12   Parabolic dish of the Arecibo radio telescope Arecibo National Laboratory in Puerto Rico. The dish diameter is 305 meters or 1000 feet.

This leads to one of the limiting parameters of parabolic dish arrays. Surface irregularities, either from manufacturing tolerances, or in this case, possible "frost heaves," result in less than perfect phase relationships among the reflected waves. Perhaps this is why Arecibo was built in Puerto Rico rather than frigid New England!

Feeding Antennas for Space Communication

Feeding Antennas for Space Communication

Space communication is generally limited by the receive signal-to-noise ratio, since transmitter power in satellites, or distant terrestrial bodies, is usually in short supply. With an antenna pointed toward space, unless the sun crosses the path, the antenna is seeing its signal sources against a cold distant sky. Unfortunately, with the feed systems we have discussed, any sidelobes or spillover from the feed that misses the reflector heads towards earth. This is not a significant problem for the transmitted, or uplink, system — just turn up the transmit power a notch.

On the receive side, it is a whole different story. Random noise signals received from the warm earth are

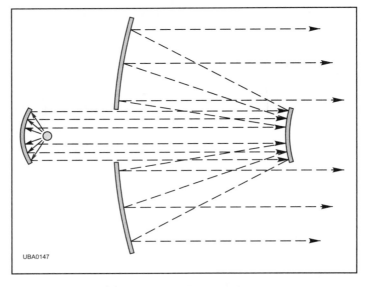

Fig 16-13 — Illustration of the Cassegrain feed system — adapted from optical telescope technology.

much more powerful than those from space and can be the limiting factor in signal-to-noise ratio. If the pattern from the feed is tightened to stay entirely within the reflector, it will undoubtedly not fill it completely, thus giving up some of the possible forward gain.

One alternative is to feed the antenna from the rear using a system adapted from optical telescopes called a *Cassegrain* feed. This is illustrated in **Fig 16-13**. Note that with this feed system that if any radiation misses the sub-reflector at the main dish focus point, it heads for space rather than towards the warm earth.

The Cassegrain feed is a particularly good choice for large reflectors, since a long transmission line to the feed horn is eliminated. A low-noise receiver amplifier at the feed horn can be any size without blocking the pattern and is also in a location where it is easy to service, rather than being perhaps 100 feet over the top of the dish.

Chapter Summary

This chapter has discussed the use of surface-type reflectors used to provide gain and directivity to antenna systems. This is one alternative to the multielement arrays of driven elements I discussed in previous chapters. I will discuss other possibilities later in this book. Of the types discussed, the parabolic dish is almost universally encountered at microwave frequencies, whether used for TV satellite receive dishes, search radar antennas or point-to-point communications links.

Review Questions

16-1. Discuss the relative benefits of unidirectional phased arrays compared to surface-reflector antennas. Under what conditions might each have the edge?

16-2. Can you think of two "structures of opportunity" that could be pressed into service as plane reflectors?

16-3. Examine the two-element horizontal broadside array of Fig 7-1 and its gain shown in Fig 7-5. If you were to put a properly sized plane reflector $\lambda/4$ behind it, what would you expect the forward gain to be?

16-4. Calculate the receive aperture in wavelengths of a 15 foot parabolic dish at 2300 MHz and compare it to that of a 6 foot dish at 5600 MHz. Compare the gains of each in Table 16-1. What do you conclude?

Chapter 17

Surface Reflector Antennas You Can Build

This reflector antenna for 70 cm satellite work can be built from parts available at any hardware store.

Contents

Surface Antennas You Can Build

There are a number of easy to build surface reflector arrays that can be constructed in the home shop and provide excellent performance at low cost. This chapter will provide some representative examples; however, availability of materials can make some other configurations equally attractive.

A Simple UHF Plane Reflector Array

This antenna may become a classic! It provides exactly the performance needed for successful low earth orbit (LEO) satellite operation, costs relatively little and is easy to duplicate. The antenna was first presented in *QST* Magazine, but is worth repeating here.[1]

The antenna in **Fig 17-1**, consists of two broadside dipole pairs, one horizontal and one vertical, in front of a 2 foot square reflector (a 2½ foot square would work a bit better, but would be a bit more cumbersome). For terrestrial use, either polarization may be used. By feeding both polarizations with a 90° phase delay, circular polarization can be provided, which provides better communication to either a tumbling satellite using linear polarization or to a satellite that uses circular polarization in the same sense.

Fig 17-2 shows the configuration layout, while **Fig 17-3** shows the hardware store perforated aluminum sheet reflector with the dipoles in place. The reflector is made rigid by pieces of 1×1×⅛ inch aluminum angle around the edges and 1×⅛ inch bar stock across the dipole mounting locations. Each dipole is made from two 5½ inch long pieces of ¾ inch OD aluminum tubing insulated at the center with ½ inch black PVC lawn sprinkler couplings. The dipoles are insulated from, and mounted on, aluminum angle brackets that provide a separation of 5⁵⁄₁₆ inches from the reflector and 19½ inches from its opposite number.

Fig 17-1 — View of a 70 cm panel reflector antenna. Its compact size is evident.

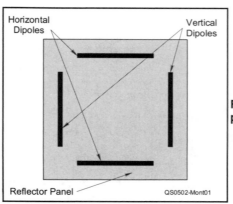

Fig 17-2 — Layout of panel reflector.

The feed arrangement for right-hand circular polarization is shown in **Fig 17-4**. The cables to each pair of dipoles are each λ/2 (electrical length) longer on one side of a pair than another, so the connections to the dipoles are reversed to make the signals of each pair be the same phase. In addition, the cables to the horizontal pair are 90° longer, resulting in generation of right-hand circular polarization.

The "four-way power splitter" in the center consists of a pair of T adapters to join the two 50 Ω loads from each of the pair of dipoles. The resultant impedance of the two in parallel is 25 Ω. Each of the 25 Ω loads is transformed through a λ/4 length of 50 Ω coax to 100 Ω and then each is combined in parallel in another coaxial T to form a 50 Ω load for the transmission line to the radio equipment. The author used 50 Ω Type N connectors and Ts for all connections. While these are somewhat more expensive than the UHF type, they are recommended for use at VHF and above because they offer a constant impedance. In addition they are waterproof if properly assembled.

Additional details for construction can be found in the *QST* article or on the ARRL Web page.[2]

Fig 17-3 — View of panel reflector with dipoles in place.

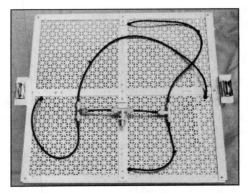

Fig 17-4 — Detail of feed cables and four-way power divider.

High Performance VHF or UHF Array

Many serious VHF and UHF operators interested in long-haul communications have selected arrays of Yagi arrays (Yagis are described in a subsequent chapter) for their stations. Still, for those without a suitable method of confirming the Yagi's correct tuning and spacing, the plane reflector can provide a predictable alternative that is easy to duplicate in the home workshop.

The antenna shown in **Fig 17-5** brings together material from a number of chapters resulting in a combination broadside-collinear array in front of a reflecting backscreen. The dimensions given are for a design frequency of 144.5 MHz, but nothing is critical about the dimensions, nor

the design frequency. This makes duplication straightforward. The *EZNEC* predicted gain of more than 18 dBi (see **Fig 17-6**) is quite impressive and hard to beat with homemade Yagis.

Element Construction

The antenna is composed of three pairs of center-fed full waves (or perhaps better yet, collinear end-fed half waves) stacked vertically to give good gain along the horizon. Each full wave

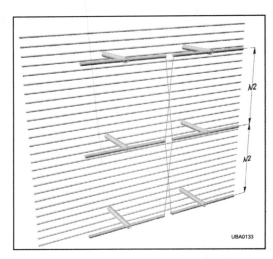

Fig 17-5 — Configuration of combination six element collinear-broadside array in front of screen reflector.

element should be about 71 inches tip to tip, with a gap between the inside ends for transmission-line connection. The upper and lower collinear elements are connected to the center pair through readily available 450 Ω ladder line from the center pair to those above and below.

Elements are made from ⅜ inch OD aluminum tubing with each λ/2 piece about 35 inches long (a bit shorter than the usual λ/2 because of the increased diameter) and spaced 20 inches from the backscreen. The elements should be supported at their center, a zero-voltage point, but it is best to use nonconducting material, such as PVC tubing, to be safe. Connections to the rods at their common ends can be made by flattening the end of each tube and drilling each for bolts, lockwashers and nuts to secure ring terminals soldered to the end of the transmission line sections.

The diameter and spacing of the driven elements results in an appropriate impedance that divides the power almost equally in thirds between the elements when the array is driven from the center pair. This results in maximum forward gain. It is important that the lines between the elements be close to an electrical λ/2 and twisted one turn, as shown in Fig 17-5, or alternately 1 λ long and not twisted, to obtain the correct phase in all three elements.

Feed Arrangement

The resultant impedance results in equal power in each pair and a good match for low-loss 300 Ω feed line going back to the station. As shown in **Fig 17-7**, the relatively flat curve of SWR vs frequency confirms that this antenna is not fussy about precise dimensions.

If coax cable is desired instead of 300 Ω line, a simple λ/2, 4:1 loop balun can be used to make the transition to readily available 75 Ω cable TV (CATV) type coax for the run to the station. This coax is available in different thicknesses, with the larger sizes offering very low loss compared to most types of coax used by amateurs. Cable is often available from CATV installers at no cost, as they often have end pieces on their reels that they can't use. The resulting 1.5:1 SWR at the radio is not generally a problem for the equipment. If a perfect match is desired, a 61 Ω, λ/4 matching section can be fabricated from copper pipe and tubing.

Backscreen

The backscreen is composed of 1.5 λ long wires that are parallel to the elements. These wires extend λ/4 above and below the driven elements. The wire lengths are not critical and any convenient construction techniques can be used. Aluminum ground wire or bare (or insulated) house wire are suitable material. Each backscreen wire is spaced 0.05 λ, about 4 inches for 2 meters, from the next. Alternately, agricultural mesh (chicken wire) can be used if its wire spacing is similar or smaller and its crossings are welded or soldered. Chicken wire should be suspended on an appropriate wood or aluminum support frame that can have attachments for the element supports.

Other Bands

The antenna can easily be scaled to operate on other bands. A second band's driven elements can be mounted on the rear of the backscreen, if the backscreen wire spacing is close enough for the higher band. If chicken wire or other two-dimensional grid wire is used, you could have a vertically polarized antenna on one side and a horizontal polarized one on the other of the reflector.

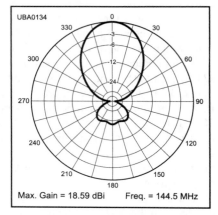

Fig 17-6 — Azimuth pattern of array in Fig 17-5 over typical ground. Note high forward gain.

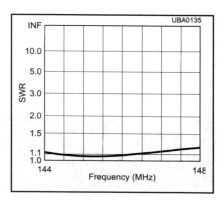

Fig 17-7 — 300-Ω SWR plot of array in Fig 17-5. Note wide bandwidth for low SWR.

Parabolic Dish Reflector Systems

While it is difficult, though by no means impossible, for the home constructor to build a parabolic dish — particularly from wire mesh material — it may be far easier to obtain a surplus dish from military, commercial or personal receive-only TV use that you can adapt to your needs. Such a dish, a surplus C band (2 to 4 GHz) receive-only TV antenna, is shown in **Fig 17-8**. Most dishes can operate efficiently at any frequency where they are at least a few wavelengths across, with larger sizes providing more gain, generally proportional to the area intercepted. The upper frequency limit of dishes is the frequency where surface irregularities or dimensional tolerances become a measurable fraction of a wavelength.

Finding the Focus

The first question upon obtaining a dish is to determine the key dimensional parameters, particularly the focus distance. Refer to **Fig 17-9**. The parameters shown can be easily measured and the following formula can determine the spacing from the center of a symmetrical dish (not those that have a distorted shape, designed for off-center feed), to the focus point.

$$f = D^2/(16\,S)$$

Once the focus point is located, you must design a feed system to illuminate the reflector from that focus point.

Designing a Feed System

The feed system is a critical element of any such reflector array. To obtain the full benefit of the dish, the beamwidth of the feed antenna must be such that most of its radiation is kept within the dish, and it must fill the dish out to its edges. If the feed only illuminates part of the reflector, the remainder might just as well not be there. On the other hand, if the feed beamwidth extends beyond the width of the reflector, the power that misses the reflector is equivalent to attenuating that amount of power. Once you've found the focus location, as described above, finding the target beamwidth of the feed is easy. Look at Fig 17-9 again and note that the required horizontal and vertical beamwidth (BW) can be found as follows:

$$BW = 2\,\tan^{-1}((D/2)/f)$$

In addition, the feed itself must be sufficiently small that it does not block a significant portion of the reflector surface. The fraction of the surface in the shadow of the feed, as well as its support structure, reduces the effective area of the dish and thus acts as an additional attenuator.

With all of those constraints in mind, virtually any relatively small unidirectional array can be used as a dish feed.

Fig 17-8 — A C band television receive-only dish modified for amateur space communication with motor-driven azimuth and elevation controls.

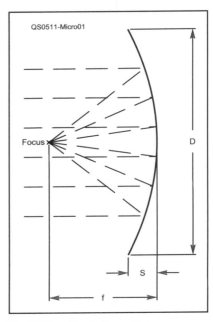

Fig 17-9 — Technique for determining focal point of symmetrical parabolic dish.

The only ones I've discussed to this point are cardioid arrays, which have a very wide beamwidth compared to most parabolas, and also plane or corner reflectors, which leave a relatively large shadow because of their size. Subsequent chapters will discuss various other options better suited for this service. These include horn, patch and Yagi antennas.

Summary

This chapter has discussed the construction of various types of unidirectional arrays you can construct using non-resonant surface reflectors. They generally operate with the simplicity of an optical mirror and, as a consequence, avoid many of the problems associated with the driving of phased arrays. Plane reflectors have been successfully employed from HF through UHF, while parabolic dishes are often the best choice in the microwave region and above.

Notes

[1]A. Monteiro, AA2TX, "A Panel-Reflector Antenna for 70 cm," *QST*, Feb 2005, pp 36-39.

[2]**www.arrl.org/files/qst-binaries/panel-antenna.zip**.

Review Questions

17-1. Discuss the relative merits of unidirectional phased arrays compared to surface-reflector antennas. Under what conditions might each have an edge in performance?

17-2. What are some of the limitations of parabolic-dish arrays? How can such limitations be overcome?

17-3. What would be the effect of using a backscreen of vertical wires behind a horizontally polarized antenna?

Chapter 18

Antenna Arrays With Parasitically Coupled Elements

An extreme example of parasitically coupled antennas. This is an array of 64 15 element Yagis.

Contents

Antenna Arrays With Parasitically Coupled Elements

All the multi-element antenna arrays I have discussed so far have had some portion of the transmitter power feeding each of the elements directly through transmission lines. I have also discussed the fact that nearby elements couple to other elements via electromagnetic fields. In addition to having an impact on the element impedances, such mutual coupling results in changes in the directional patterns in the reception and transmission of signals. Now, I turn to arrays that purposely use mutual coupling.

An Array With One Driven and One Parasitic Element

A *parasitic* antenna system is one in which one or more elements are coupled to the driven element by mutual coupling only. The first example of this is shown in **Fig 18-1**. Here I take my typical resonant λ/2 dipole at 10 MHz, λ/2 above typical ground, and place another dipole of the same length — but with no transmission line attached to it — λ/4 behind it at the same height. This second dipole is just a λ/2 length of wire, with no connection to anything else.

This type of parasitic array using a driven element and other parasitic elements is named the *Yagi-Uda* array, in honor of the two Japanese scientists who developed it (although the array's name is often abbreviated to just *Yagi*). More complex Yagi arrays than the simple two element one in Fig 18-1 will be discussed in more detail in the next chapter.

Parasitic Element Impact on Pattern

A dipole by itself has a bidirectional pattern, as shown in **Fig 18-2**. Note that the peak gain after ground reflection is 7.4 dBi. By adding a single parasitic element, the pattern shifts to a *unidirectional* one. See the elevation plot in **Fig 18-3** and the azimuth plot in **Fig 8-4**. The gain in the direction opposite the side with the parasitic element is more than 3 dB higher than the dipole, while the front-to-back ratio is almost 10 dB.

It is interesting to compare this two

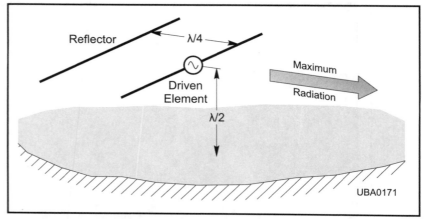

Fig 18-1 Configuration of dipole with driven element and one parasitic element behind it.

element parasitic array to the cardioid unidirectional array presented earlier. Both arrays could be made from dipoles or vertical monopoles. While the pattern of a cardioid array is more dramatic than the pattern for the two element parasitic array in Fig 8-4, the deep null at the back exhibited by the cardioid only comes about from careful design and construction of a feed system that ensures proper phase shift and equal currents in each element. By contrast, the simplicity of the simple 2-element parasitic array is hard to beat.

Parasitic Element Impact on Dipole Impedance

In addition to the impact on the radiation pattern, a parasitically

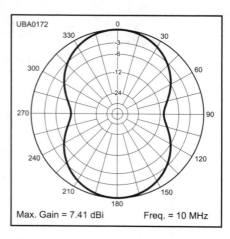

Fig 18-2 — Azimuth pattern of reference dipole.

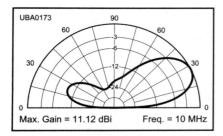

Fig 18-3 — Elevation pattern of antenna in Fig 18-1.

coupled element will also have an impact on the feed-point impedance of the driven element. **Fig 18-5** shows the modeled SWR curve of a reference dipole at a height of λ/2. By adding a λ/2 long parasitic element λ/4 away, the resonant frequency shifts downward more than 2%, as shown in **Fig 18-6**. Fortunately, you can compensate for this frequency shift by shortening the driven dipole to make it resonant again in the presence of mutual coupling to the second element. It becomes easier to feed the driven element using a typical coax line, while the unidirectional radiation pattern remains essentially unchanged. In other words, tuning the driven element alone has no effect on the radiation pattern.

What About Other Spacings?

I started with a spacing of λ/4, mainly because of the physical similarity to the driven cardioid configuration I described earlier in Chapter 7. There is nothing magic about this spacing distance, and you can obtain interesting results using

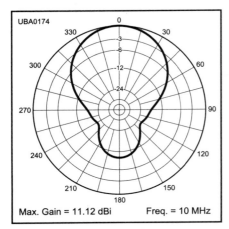

Fig 18-4 — Azimuth pattern of antenna in Fig 18-1.

other spacings. As you move the λ/2 long parasitic element closer than λ/4 to the driven element, the general trend is a lowering of the feed-point impedance of the driven element, plus a reduction in gain and front-to back ratio.

An interesting phenomenon occurs as you reach very close spacings with a λ/2 long parasitic element. At a spacing of 0.085 λ, for example, the front-to-back ratio approaches unity and you'll see the azimuth pattern shown in **Fig 18-7**. While this shares the generally bidirectional pattern of a single dipole, the beamwidth is significantly narrower and the maximum is almost 3 dB higher than the dipole by itself. A look at the predicted SWR curve indicates that the driven element impedance drops to about 9 Ω. This is undesirable, since losses in the system begin to become an appreciable fraction of this small resistance.

Since you've passed the bidirectional point at a 0.085 λ spacing, as you move even closer you probably won't be surprised to find that the pattern now reverses itself. At a spacing of 0.05 λ, for example, you get the pattern shown in **Fig 18-8** with a λ/2 long parasitic element. At this close spacing the driven-element impedance has dropped to around 4 Ω, which makes for an even more difficult, and lossy, match.

Here's some terminology you will often encounter in discussions about parasitic arrays. If a parasitic element results in radiation away from it towards the driven element, it is called a *reflector*. On the other hand, if the radiation becomes focused in the direction moving towards the parasitic element from the driven element, the parasitic element is called a *director*.

Going to wider spacings affects things less dramatically than narrower spacings. The mutual coupling is reduced and the parasitic element has less effect on pattern and feed-point impedance. Both the gain and front-to-back ratio are generally reduced, but the driven element impedance rises. At a spacing of λ/2, the pattern actually has started to broaden, so there may be some applications for which a wide pattern

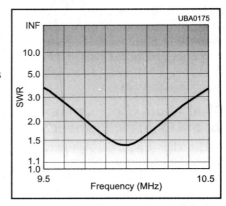

Fig 18-5 — SWR of single λ/2 dipole λ/2 above typical ground.

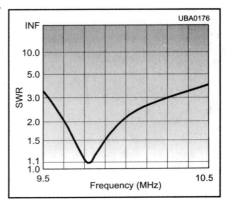

Fig 18-6 — SWR of driven λ/2 long dipole element with same-length parasitic element spaced λ/4 to the rear of driven element, both λ/2 above typical ground.

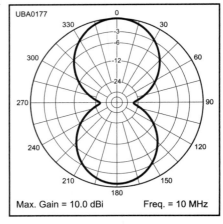

Fig 18-7 — Azimuth pattern of array with same-length parasitic element spaced 0.085 λ to the rear of driven element, both λ/2 above typical ground.

like that in **Fig 18-9** would be useful. The resonant impedance is back to about 80 Ω; however, the resonant frequency has shifted upwards some. At even wider spacings, there is little of interest to discuss.

What About Other Lengths for the Parasitic Element?

The other key parameter you can adjust is the length of the parasitic element — in other words, you can tune it. In an attempt to optimize forward gain and front-to-back ratio, I adjusted spacing and length of the reflector until I obtained the result shown in **Fig 18-10**. This is at a spacing of 0.2 λ with the reflector about 4% longer than the driven element. As a side note, it is an awful lot easier to go through these studies with an antenna modeling program such as *EZNEC* than to go out in the back yard, lower the antenna, change a parameter, hoist up the antenna and take readings — again and again and again!

It is also possible to head in the other direction, with closer spacing and a shorter parasitic element. By moving to a spacing of 0.085 λ and reducing the length of the parasitic element by 2.5% compared to the driven element (turning it into a director), the pattern of **Fig 18-11** was obtained. At this spacing and length, the driven element has an impedance of about 23 Ω.

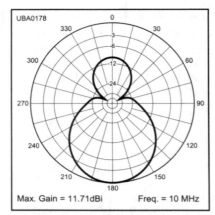

Fig 18-8 — Azimuth pattern of array with same-length parasitic element spaced 0.05 λ to the rear of driven element, both λ/2 above typical ground.

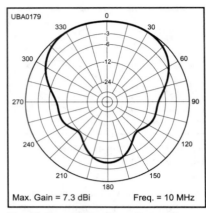

Fig 18-9 — Azimuth pattern of array with same-length parasitic element spaced 0.5 λ to the driven element, both λ/2 above typical ground.

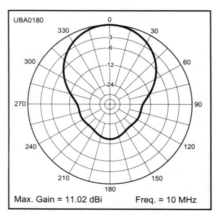

Fig 18-10 — Azimuth pattern of array with tuned parasitic reflector spaced 0.2 λ to the rear of driven element, both λ/2 above typical ground.

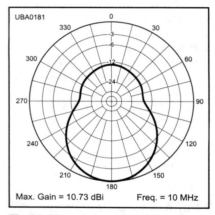

Fig 18-11 — Azimuth pattern of array with tuned parasitic director spaced 0.085 λ from the driven element, both λ/2 above typical ground.

An Array with One Driven and Two Parasitic Elements

Since a single parasitic element in a two element Yagi parasitic array can provide such benefit with relatively little effort, you may wonder whether adding more parasitic elements might do even more. The answer is yes.

The next chapter will go into more detail about such arrays, but the addition of a reflector and a director to the driven element (making a three element Yagi) can provide a good introduction to that discussion. The resultant pattern of a reflector at 0.2 λ spacing behind the driven element and a director at 0.085 λ ahead of the driven element is shown in **Fig 18-12**. The front-to-back ratio is now almost 25 dB, impressive indeed.

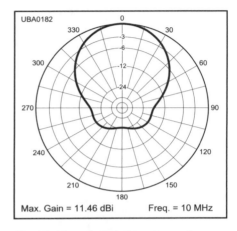

Max. Gain = 11.46 dBi   Freq. = 10 MHz

Fig 18-12 — Azimuth pattern of 3-element Yagi array with parasitic reflector and director.

Summary

I have introduced the concept of the use of parasitic (not-connected) elements in antenna arrays. These elements allow beam shaping without the problems involved obtaining correct phase relationships in previously described driven arrays. Neither does the parasitic array have the penalty of the large wind loads common to surface-reflector arrays.

Review Questions

18-1. Under what conditions might an array with one or more parasitic elements have advantages over directly driven phased arrays or surface reflector arrays previously described?

18-2. What is a negative consequence of closely coupled parasitic elements?

18-3. What can be the effect of cutting parasitic elements with a few percent error? How does this compare with driven arrays?

Chapter 19

The Yagi-Uda or Yagi, Parasitically Coupled Antenna

Yagis for multiple bands surround a higher frequency parabolic reflector array.

Contents

The Yagi-Uda, or Yagi, Parasitically Coupled Antenna

The last chapter discussed the concept of parasitically coupled antennas. This antenna configuration was extensively evaluated and developed by two Japanese academics in the early part of the last century.

The concepts of the antenna later known as the *Yagi-Uda* were published in 1926 by Shintaro Uda, with the collaboration of Hidetsugu Yagi, both of Tohoku Imperial University, Sendai, Japan. Yagi published the first English-language reference on the antenna in a 1928 article and it came to be associated with his name.[1] However, Yagi always acknowledged Uda's principal contribution to the design, and the proper name for the antenna is the Yagi-Uda antenna (or array).

The Yagi was first widely used by Allied Forces as the antenna for VHF and UHF airborne radar sets during WW II, because of its simplicity and directionality. The Japanese military authorities first became aware of this technology after the Battle of Singapore, when they captured the notes of a British radar technician that mentioned "Yagi antenna." Japanese intelligence officers did not even recognize that Yagi was a Japanese name in this context. When questioned the technician said it was an antenna named after a Japanese professor. (This story is analogous to the story of American intelligence officers interrogating German rocket scientists and finding out that Robert Goddard was the real pioneer of rocket technology even though he was not well known in the US at that time.)[2]

Current usage in the US is to refer to this antennas configuration as a *Yagi*, and we will use that term in this book, with the understanding that Yagi is really short for "Yagi-Uda."

What's a Yagi All About?

The antenna developed by Uda and Yagi took the basic three element parasitic array described in the last chapter and extended it, through the addition of multiple directors, to provide a highly focused beamwidth signal for VHF and UHF (the "ultra short waves" in Yagi's article title). They empirically developed the early designs of such arrays, including the optimum lengths and spacings. The results have been relied on for many years, until the availability of modern computer modeling and simulation.

Why More Directors and not More Reflectors?

An effective reflector results in minimal signals behind the array. Thus additional reflector elements placed *behind* the first reflector receive only a very small residual signal going past the reflector element and therefore contribute little to the signal leaving at the front of the array. Occasionally, a multielement reflector structure can be found, but this is really just a different geometry at the approximate distance of the primary reflector. Sometimes a plane reflector is used behind the driven element for better front-to-back performance.

Additional directors, on the other hand, *are* in the path of the main beam formed from the driven element, reflector and any previous directors. Thus they continue to focus the beam further as it progresses forward along the antenna axis.

What Frequencies are Appropriate for Yagi Antennas?

While the original paper about Yagi-Uda antennas was focused (no pun intended) on VHF and UHF applications, there is nothing about that range that is specific to the Yagi design. On the low frequency side, the limits are the physical size of the structure. On the high frequency side, manufacturing tolerances, plus the effectiveness of other available alternatives, form the limits for the deployment of Yagis.

At upper HF and into the UHF range, Yagis tend to be made from rigid, self-supporting materials, such as aluminum tubing. At the low end, they are more often found as fixed antennas constructed of wire elements. At the upper end of the frequency range, they are sometimes even seen etched onto printed circuit material.

Enough Talk; Let's See Some Examples!

A discussion of Yagi antennas would probably start with the three element parasitic array I introduced at the end of the last chapter. However, Yagi development really gets into full swing with antennas having multiple directors. In keeping with the spirit of the original Yagi paper, I will shift the design frequency into the VHF range, but keep in mind that the results can be scaled to any frequency.

At VHF, ground reflections at any reasonable height rapidly shift through peaks and nulls. Thus I will change the usual environment to that of *free space* when generating patterns at these frequencies. **Fig 19-1** illustrates the point. Here you have the elevation patterns of identical 145 MHz reference dipoles, constructed of the same 0.25 inch aluminum tubing that I will use for the Yagis to come. One antenna is mounted 40 feet above real ground; the other is mounted in free space. The rapid fluctuations of field strength over ground tend to distract from the comparison between antenna designs. You will need to keep in mind if comparing antenna results that the main lobe gain of a free-space dipole is 2.13 dBi, while the gain of a dipole at the peak of the first elevation lobe over typical ground is 7.67 dBi. The difference of 5.54 dB is due to ground reflection gain and is close to the theoretical limit of 6 dBi. The ground reflection gain should be carried over when evaluating other free space designs.

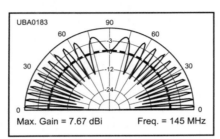

Fig 19-1 — **Comparison between elevation patterns of 145 MHz reference dipole in free space (dashed line) and the same dipole mounted 40 feet above typical ground (solid line).**

The Medium Sized Three Element VHF Yagi

As noted, the three element Yagi is just an optimized and further developed version of the array with reflector and director described in Chapter 18. The configuration is shown in **Fig 19-2**. The spacing between the elements of a Yagi is a key design parameter. In general, wider spacings provide higher performance; while closer spacings result in a more compact physical configuration, but with less gain. You usually select a boom (the structural element supporting the elements) and then optimize the number of elements and their spacing on the available boom length to maximize performance. For my first example, I will start with a boom length of 0.3 λ and use 0.15 λ spacing between the driven element and both the reflector and director.

Azimuth Pattern

Careful adjustment of element lengths to obtain a resonant driven element, maximum front-to-back ratio and clean forward pattern with acceptable forward gain resulted in the element lengths shown in **Table 19-1**. The resulting free-space azimuth pattern is shown in **Fig 19-3**. The forward free-space gain of 7.84 dBi corresponds to a gain (compared to the gain of 2.13 dBi of a dipole in free

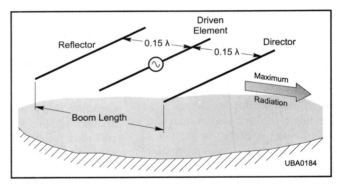

Fig 19-2 — **Configuration of three element Yagi.**

Table 19-1

Dimensions of the 3-Element Yagi with 0.15 λ Reflector and Director Spacing. All Elements 0.25 inches in Diameter.

Element	Length (λ)	Spacing from DE (λ)	Length (inches)	Spacing from DE (inches)	Difference (%)
Relector	0.490	0.15	40.53	12.2	3.8
Driven Element	0.472		38.42		
Director	0.440	0.15	35.82	12.2	−6.8

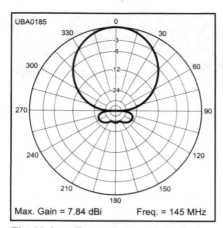

Fig 19-3 — Free-space azimuth pattern of three element Yagi tuned for front-to-back ratio (F/B).

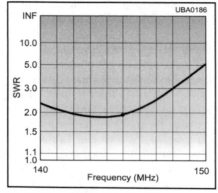

Fig 19-4 — SWR (50 Ω) of three element Yagi in Fig 19-2.

space) of 5.7 dB. This is almost four times the power in the peak of the main lobe of a dipole in free space.

A key issue in the design of Yagi arrays is that the element spacings and lengths can be adjusted to obtain maximum front-to-back ratio, maximum forward gain or match to a particular impedance — but not all, or even two, parameters can be optimized using the same dimensions. It is thus important to decide what your design goals and acceptable parameter limits are before you start.

Matching to a Transmission Line

The 50 Ω SWR for the example three element Yagi is shown in **Fig 19-4**. Note that while the frequency of lowest SWR is somewhat below the design frequency of 145 MHz, the actual resonant frequency, the

frequency with 0 Ω reactance, is at 145 MHz. The impedance at resonance is about 26 Ω, which is not a very convenient impedance for use with common transmission lines. This is fairly typical of Yagi antennas and I will discuss some of the techniques used to match transmission line to the generally low impedance of Yagi driven elements a bit later.

In general, this is a solved problem and I will not dwell on it as I talk about the Yagi properties in this section. For example, in this case, a λ/4 section of 36 Ω transmission line (which is available, but two common 75 Ω lines in parallel could also be used) will transform the 26 Ω to 50 Ω. The SWR plot transformed to 26 Ω, under the assumption that some sort of matching arrangement is provided is shown in **Fig 19-5**.

Yagi Operating Bandwidth

For many antenna types, the key parameter determining operating bandwidth is the frequency range over which a usable SWR is provided. Depending on requirements, that definition may or may not be appropriate for a Yagi. For example, the 26 Ω 2:1 bandwidth of the three element Yagi in Fig 19-2 is from around 141 to 147.5 MHz, as shown in **Fig 19-5**. This is about a 4.5% range around the center frequency, compared with around 8% for a similar diameter dipole.

As the frequency is changed within the operating bandwidth of a dipole, however, the pattern remains virtually constant. This is not the case with a Yagi. **Fig 19-6** shows the azimuth pattern of the Yagi in Fig 19-2 at both ends of the SWR bandwidth. Note that the forward gain is actually a bit higher, but the front-to-back ratio is significantly reduced at each end of this frequency range. Had I optimized for forward gain, rather than front-to-back ratio, you would have seen a different form of change. Whether the difference from design center performance is significant depends on your application.

It is often possible to adjust the element lengths to result in slightly less than optimum performance at a spot

frequency, but for more satisfactory performance over a wider bandwidth. This is often done by making the reflector a bit longer and the director a bit shorter than optimum.

Changing the Element Spacing

Within the constraint of a boom length of 0.3 λ, you can distribute the element spacings differently and come up with a different antenna. One choice would be to have the reflector 0.2 λ behind the driven element and the director 0.1 λ in front. This is the kind of change Uda and Yagi made empirically in the laboratory, but it is even easier to do with a modeling program like *EZNEC*!

The result, perhaps with a bit of optimization left for the student, is shown in **Fig 19-7**. And **Fig 19-8** shows the azimuth with the spac-

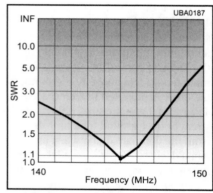

Fig 19-5 — SWR (26 Ω) of three element Yagi in Fig 19-2.

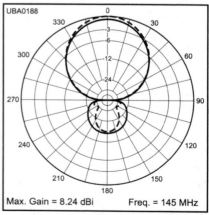

Fig 19-6 — Azimuth pattern of three element Yagi in Fig 19-2 at 2:1 SWR limits.

ings reversed — the reflector at 0.1 λ behind the driven element and the director 0.2 λ in front. The forward gain is about 1 dB higher in this arrangement and the front-to-back ratio has improved slightly. The feed-point impedance is, however, reduced to about 10 Ω following the tweaking of lengths to regain a reasonable front-to-back ratio. There is no free lunch!

The Short and Long of Three Element VHF Yagis

If space constraints are paramount you can employ shorter boom lengths, at some cost of performance. **Fig 19-9** shows an example of a three element Yagi on a 0.1 λ boom, with equally spaced director and reflector. As is evident, this design gives up from 1 to 2 dB of forward gain compared to previous antennas with three times the boom length. The front-to-back ratio is also reduced compared to the larger antennas. This antenna has a resonant feed-point impedance of about 11 Ω.

You can also go in the other direction, using a longer boom and wider element spacing. If you make the array roughly square at 0.5 λ per side with a boom of that length, you can have a bit more of everything. I will later discuss that such a boom may be better served by adding a director or two, but the three element design can be close in performance and is lighter to raise and turn. It also has less wind loading. With careful optimization, this antenna can have additional gain, but probably not as good a front-to-back ratio as the 0.3 λ design presented earlier. It also can provide a direct 1:1 match to 50 Ω coax cable.

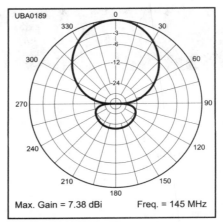

Max. Gain = 7.38 dBi   Freq. = 145 MHz

Fig 19-7 — Azimuth pattern of 0.3 λ boom three element Yagi with 0.2 λ reflector and 0.1 λ director spacing.

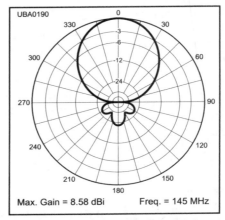

Max. Gain = 8.58 dBi   Freq. = 145 MHz

Fig 19-8 — Azimuth pattern of 0.3 λ boom three element Yagi with 0.1 λ reflector and 0.2 λ director spacing.

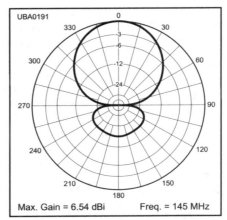

Max. Gain = 6.54 dBi   Freq. = 145 MHz

Fig 19-9 — Azimuth pattern of 0.1 λ boom three element Yagi with 0.05 λ reflector and director spacings.

Yagis with More Elements

What I've discussed so far has set the stage for more interesting designs of long-boom Yagis with more elements — the focus of the original Yagi-Uda work. The design variables and adjustable parameters are the same — boom length, the number of elements, element spacings and element lengths. The outcomes are the same as well — forward gain, front-to-back ratio, driving impedance and operating bandwidth. As for the earlier designs, the optimum for all parameters never occurs with any one design.

With each element you add to the mix, you get two independent variables — the element's length and its spacing from neighboring elements. Since the number of combinations of adjustments goes up with the square of the variables, it can become quite a challenge to be sure you have found the optimum. Let's see — if I make this element a bit longer — and then adjust all the other spacings and lengths to compensate — am I getting better or worse performance? You get the idea!

Fortunately, there has been a lot of work done on this subject since Yagi's 1928 paper and you don't have to reinvent every wheel. *The ARRL Antenna Book*, for example, lists lengths and spacings for many proven, optimized Yagi designs for the HF and VHF/UHF amateur bands. These designs can be scaled to provide a starting point for other applications or frequency bands too.

The Boom Length — Element Count Issue

When you start looking at antennas with boom lengths longer than about λ/3, you sometimes run into an interesting marketing gambit. Antenna manufacturers tend to market their long Yagis based on element count, with the boom length buried in the specifications sheet. In fact, boom length is arguably the most critical parameter in long boom Yagi design. For any given boom length, there is an optimum design (including an optimum number of elements) that will provide the gain shown in **Fig 19-10**.

You can also obtain the same gain with an antenna on a given boom length with more elements; however, that doesn't mean that this design has more value than one with less elements — in fact, since most antenna installations are designed to withstand a certain level of wind force, having more directors (and hence more wind-surface area) may well be a disadvantage. Watch this if you are considering buying an antenna array.

A Short "Long Yagi"

The azimuth plot of an example of an array with a boom length of 1 λ is shown in **Fig 19-11**. Note that the forward gain is close to that predicted in Fig 19-10. Its dimensions are provided in **Table 19-2** for the case with a nonmetallic boom. Metal booms may also be used; however, an adjustment in element length is usually required for them.

A Longer "Long Yagi"

Extending the Yagi to a boom length of a bit less than 2 λ results in the azimuth pattern of **Fig 19-12**. Again the forward gain is close to that predicted in Fig 19-10. This analysis could be extended as far as desired; however, a few observations are in order.

If you were to place a six element combination broadside-collinear

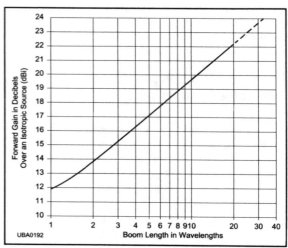

Fig 19-10 — Gain of optimized Yagi as a function of boom length (Courtesy of *The ARRL Antenna Book*).

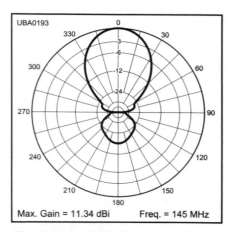

Fig 19-11 — Azimuth pattern of seven element Yagi on 1 λ boom.

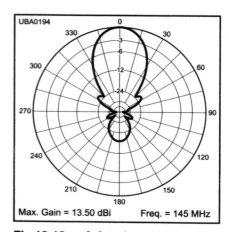

Fig 19-12 — Azimuth pattern of 10 element Yagi array on 2 λ boom.

Chapter Summary

Table 19-2

Dimensions of the seven element Yagi on 1 λ Insulated Boom. All Elements 0.25 inches in Diameter.

Element	Length (λ)	Spacing from R (λ)	Length (inches)	Spacing from R (inches)	Difference (%)
Relector	0.503	0	40.94	0	6.1
Driven Element	0.474	0.15	38.58	12.28	
Director 1	0.458	0.216	37.32	17.59	−3.3
Director 2	0.450	0.338	36.61	27.52	−5.1
Director 3	0.438	0.508	35.67	41.34	−7.5
Director 4	0.432	0.717	35.20	58.35	−8.8
Director 5	0.431	0.961	35.12	78.19	−9.0

array in front of a backscreen reflector (such as was shown earlier in Fig 17-5 in Chapter 17), the resulting reflector array provides a forward gain within about 0.5 dB of the longer 2 λ Yagi array. The antenna with the backscreen reflector, however, is about 1.5 λ high, 1.5 λ wide and 0.25 λ deep, while the Yagi is 2 λ long, about 0.5 λ wide, and not very high. You're trading, in effect, volume for area for two different arrays with comparable gains.

Whatever you decide to call "optimum" for any particular installation depends on your specific requirements. I should also point out that the reflector array from Chapter 17 has much less sensitivity to dimensional tolerances or changes in frequency, while the long-boom Yagi is more sensitive to these effects.

You could keep extending the boom and gaining additional forward gain. Another alternative is to *stack* another copy of the array on top of the first. At the optimum stacking distance, you can increase the gain by almost 3 dB. Stacking compresses the elevation pattern, rather than further squeezing more gain out of the azimuth pattern, as happens when you lengthen the boom of a Yagi. Stacking is often a good method to get more gain, since doubling the boom length results in an increase of about 2 dB.

A Yagi antenna array can have significant gain and front-to-back ratio compared to other types of structures. It offers more gain per unit of wind-surface area than many other types of high-gain systems, at the cost of having tighter dimensional tolerances and correspondingly narrow operating bandwidth. Multiple Yagi arrays can often be combined into stacks of very effective antenna systems.

Notes

[1]H. Yagi, "Beam Transmission of Ultra Short Waves," Proceedings of the IRE, 16, pp 715-740, Jun 1928.

[2]**en.wikipedia.org/wiki/Yagi_antenna**.

Review Questions

19-1. What are some of the major advantages of Yagi antennas compared to other types of systems previously covered?

19-2. What are some of the major disadvantages of Yagi antennas compared to other types of systems previously covered?

19-3. Under what conditions might a stacked arrangement of Yagis be more useful than a single Yagi with the same total boom length. What are the disadvantages of such a system?

Chapter 20

Practical Yagis for HF and VHF

Yagis for two frequency ranges share a tower.

Contents

Practical Yagis for HF and VHF

The last chapter discussed the structure and results obtained with Yagi antennas. Yagis are found in antennas designed for MF through the middle of the UHF region. As has been mentioned, one downside of Yagi antennas is that satisfactory operation requires close attention to tight dimensional tolerances. I will describe here some representative, reproducible Yagi designs. You can scale them to other frequencies, but at your own risk!

Matching A Yagi to a Transmission Line

As discussed earlier, one potentially unpleasant characteristic of Yagi antennas, particularly those with close-spaced elements, is that the driven element impedance is often quite low. This is a function of parasitic element length and spacing. While it is generally possible to find parasitic element spacings and lengths that will result in a 50 Ω driven element impedance, those dimensions are not generally compact. They are also often not the ones that give best front-to-back ratio, nor the highest forward gain.

Yagi designs are compromises. Even ignoring driven element impedance, the dimensions that give best front-to-back ratio generally do not provide best forward gain. Still, once your objectives and priorities are defined, there is likely to be some Yagi configuration that meets them. By having a feed connection arrangement that works with a wide range of impedance values, you are free to optimize your trade-off between front-to-back ratio and forward gain without being constrained unduly by impedance concerns.

Before I describe particular Yagi designs, I will discuss some of the methods used to match Yagi driven elements to transmission lines.

Q Bar Matching

One of the earliest methods used to match low impedance Yagi driven elements to 50 Ω coaxial cable was

called a "Q Bar." This arrangement, shown in **Fig 20-1**, is nothing more than a λ/4 transmission-line matching section. In order to match a given resistive antenna impedance (Z_A) to any transmission-line characteristic impedance (Z_0), a λ/4 transmission-line matching section of characteristic impedance Z_L can be inserted between them. The characteristic impedance of Z_L is equal to:

$$Z_L = \sqrt{Z_0 Z_A}$$

(Eq 1)

For example, if you had an antenna with a feed-point impedance of 20 Ω and wanted to match it to 450Ω ladder line, you would need a transmission line that had a characteristic impedance of 95 Ω.

The characteristic impedance of parallel-wire transmission line in free space is found as follows:

$$Z_L = 276 \ log_{10}\left(\frac{2D}{d}\right)$$

(Eq 2)

where D is the center-to-center spacing and d is the conductor diameter.

If you had a line made from 0.25 inch diameter rods, spaced 0.275 inches center-to-center, you could achieve a Z_0 of 95 Ω, but it would make for a tight fit. You also wouldn't want to do this for a very long distance. Note that this only works

if D > d, otherwise the two conductors would be touching. This sets the lower spacing limit at 83 Ω, not far from where you are now.

A more modern adaptation of this technique would be to match to a 50 Ω line. This could not work with the Q bars, but could be done with a transmission line with a characteristic impedance of 31.6 Ω. This could be approximated by paralleling two 75 Ω lines, with a resultant impedance of 37.5 Ω.

The Folded-Dipole Match

The two wire folded dipole was shown in Fig 11-11 in Chapter 11. This antenna takes up essentially the same space as a conventional single-wire dipole, but the feed-point impedance is four times the imped-

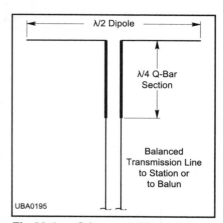

Fig 20-1 — Q-bar matching system. Note that only a limited range of values can be accommodated.

ance of a regular dipole. If a Yagi's driven element exhibits an impedance of 12.5 Ω, a folded-dipole driven element would have an impedance of 50 Ω. Note that only the driven element, not the parasitic elements, needs to be changed to a folded-dipole configuration to obtain this transformation.

The Delta Match

An *autotransformer* is a transformer with a single winding in which taps on parts of the winding provide the desired voltage-to-current ratio, or impedance ratio. The same concept applies to antennas. A half wave antenna element has a high impedance at its ends and a low impedance at the center. If you connect a transmission line between intermediate points, you can match virtually any impedance, from that of the center to that of the end-to-end impedance.

The *delta match* is perhaps the simplest implementation of this principle. It is shown in **Fig 20-2**. There is nothing magic about the dimensions, although the two connections should be equally spaced from the center to maintain balance. In use, the connection points are adjusted until a reasonable match is obtained. This will change the element tuning somewhat, so the driven element length must then be adjusted for minimum SWR and the connection points moved again, back and forth until you obtain the desired match.

This match can be used directly with a balanced transmission line, or through a balun to coax. You could use a 4:1 loop balun to match 200 Ω connection points to 50 Ω coax. See Fig 9-6 in Chapter 9. For this match, and for the next several connection methods, the center of the unbroken driven element is at zero potential, so you can connect it directly to a metallic boom if you like. If the boom and mast are well-grounded, this may have benefits in terms of lightning protection.

Typical starting points for an HF dipole connected to 600 Ω line are at a spacing between connection points of 0.12 λ for HF (0.115 λ for VHF) and a match length of 0.15 λ. The spacing should be moved inward for

lower impedances until the best match is found. The driven element will generally need to be shortened a few percent to obtain an exact match.

The T Match

Another adaptation of the autotransformer principle is called the *T match*. It differs from the delta match mainly in the connection method. The T match uses bars parallel to the antenna element to make connections to the dipole. Matching adjustment is generally provided through the adjustable position of sliding shorting bars, as shown in **Fig 20-3**. As with the delta match, the adjustments should be carried out by moving the bars the same amount on each side to maintain balance.

The T match was introduced in a *QST* article in 1940.[1] The authors (including the soon-to-be-legendary Dr John Kraus, W8JK), described a method to connect 600 Ω transmission line to a wire dipole and determined that the spacing of the shorting bars should be 24% of the dipole length for that case, with a spacing in inches of the T bars of 114/f, in MHz.

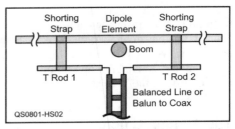

Fig 20-2 — Delta matching system.

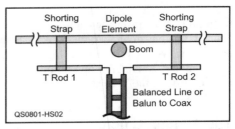

Fig 20-3 — T matching system, one of the most popular systems. It is often used with a 4:1 balun and coax feed.

For a typical Yagi made of tubing, the T-rods are usually ⅓ to ½ the diameter of the driven element. The distance between shorting bars is usually a bit less than half the driven element length.

As with the delta match, the T match is often adjusted to 200 Ω and used with a λ/2, 4:1 loop balun to match to 50 Ω coaxial cable. The T match will act to lengthen the effective dipole length. This can be accommodated by either shortening

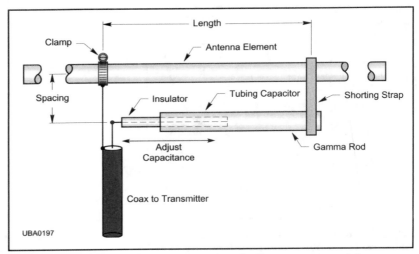

Fig 20-4 — Detail of trombone capacitor for T or gamma match.

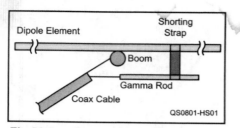

Fig 20-5 — Gamma matching system. Essentially half a T match. No balun required.

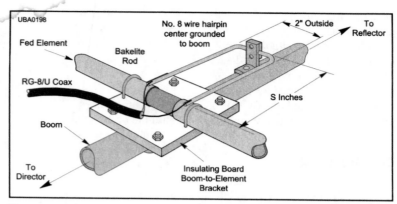

Fig 20-6 — Hairpin matching system. Can be visualized as a "folded T-match."

the dipole, or by inserting capacitance in the T bar connections. The capacitance is often made by using a "trombone" coaxial tuning section on each side of the T, as shown in **Fig 20-4**.

The Gamma Match

The *gamma match* is just half of a T match. It extends out from the boom towards one side of the driven element, as shown in **Fig 20-5**. Because it only extends to one side, just one side of the driven element needs shortening, or a trombone or other capacitance can be used to compensate for the added inductance. The gamma match, inherently unbalanced, provides for direct connection to coaxial cable without needing a balun.

For a typical Yagi made of tubing, the gamma rod is usually ⅓ to ½ the

diameter of the driven element spaced about 0.007 λ center-to-center from the driven element and about 0.04 to 0.05 λ long to match 50 Ω. A capacitance value of about 7 pF per meter of wavelength will avoid the need to shorten the driven element beyond about 3%.

The Hairpin Match

An alternative to tapping out along the driven element is to load the center of the dipole with a shunt inductance connected to each side. This brings the electrical center effectively further out along each side of the dipole. A low-loss inductor can be formed by a short (<< λ/4) section of balanced transmission line. This arrangement, shown in **Fig 20-6**, has the appearance of a hairpin and thus it is called a *hairpin match*.

Some early work, reported in *QST* in 1962, defined the hairpin parameters for a Yagi antenna operating at 14.28 MHz.[2] These are shown in **Fig 20-7** and **Fig 20-8**. The required inductance and line length (the same line width should be used) both go down directly with frequency, so results can be scaled to provide a starting point for other frequency antennas.

This kind of match is simple to construct and works well. Note that unlike most of the other techniques mentioned, it requires a split driven element. While it will work as shown in Fig 20-6, a balun should be inserted between the driving point and the coax to avoid feed-line radiation, which can distort the antenna pattern. The balun can be as simple as a coil of the coax feed line.

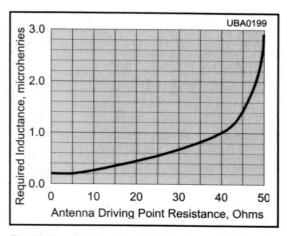

Fig 20-7 — Inductance required for hairpin matching to various driving point resistances at 14.28 MHz.

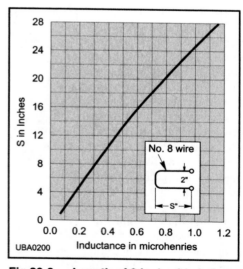

Fig 20-8 — Length of 2 inch wide hairpin inductor as a function of required inductance at 14.28 MHz.

Let's Build Some Yagis

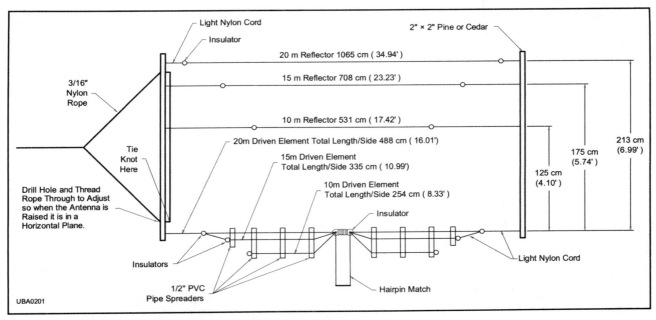

Fig 20-9 — Dimensions of the VE7CA triband 2 element Yagi.

A Portable Two Element Triband Yagi

Perhaps the simplest of Yagis is the wire triband HF Yagi designed by Marcus Hansen, VE7CA.[3] Hansen is also the designer of a wonderful homemade HF transceiver that outperforms all but the highest performance commercial HF transceivers.[4]

This antenna is made of wire elements supported between 2 × 2-inch wooden spreaders. The three bands are covered by parallel driven elements, fed with a single hairpin match. The antenna uses separate reflectors, optimally spaced for each band. While shown as a triband, two element-array, it could be easily simplified into a two element monobander, or even a three element monobander for one of the higher bands by putting the driven element near the center and adding a director element. In my experience, a three element Yagi for 20 meters would be at the outer edge of what could be asked of 2 × 2-inch spreaders, although it might work for a weekend operation, when there's no wind.

A High-Performance 144 MHz 10 Element Yagi

This antenna illustrates the use of a T match and aluminum tubing to construct an easy to duplicate, but no-compromise, VHF Yagi, first described by K1FO. This antenna is one of many cases of a basic design

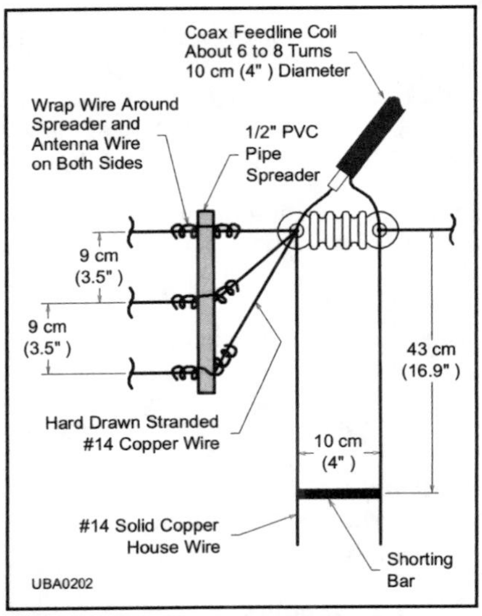

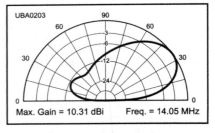

Fig 20-10 — Close-up of the VE7CA tribander's feed point.

that can be used from 10 through 19 elements described in recent editions of *The ARRL Antenna Book.*[5]

This antenna is shown in **Fig 20-13**. If you build it closely following the design specifications (given in mm to increase precision), it will yield a gain of 11.4 dBd (13.5 dBi) and a front-to-back ratio of better than 17 dB across the whole 144 to 148 MHz band. It has a 39° azimuth beamwidth when mounted as a horizontally polarized antenna.

Fig 20-11 — Predicted elevation pattern of the VE7CA tribander on 20 meters. Other bands will be similar, but for the same height (26 feet in this example) will have a peak lobe at a lower elevation angle.

The antenna is constructed on a 1¼ inch outside diameter boom with the elements spaced from the reflector as shown in **Table 20-1**. Details of the driven element construction are shown in **Fig 20-13**. The design is intended for the low end of the band, generally used for long-haul SSB and CW weak-signal work. It can be easily adapted to the FM portion centered at 147 MHz by shortening all elements by 17 mm.

The antenna is matched by a 200 Ω T-match and λ/2 long 4:1 loop balun designed to feed coaxial cable to the radio. When complete, adjust the T-match shorting bars for the best match. This should be done in its final position, if possible. If not, temporarily mount it as high as practical and point it upward to minimize near-ground reflection effects while adjusting the match.

Table 20-1

Dimensions of the 10 Element 144 MHz K1FO Yagi. All elements are ¼ inch diameter rod or tubing.

Element	Position (mm from Reflector)	Length (mm)
Reflector	0	1038
Driven Element	312	955
Director 1	447	956
Director 2	699	932
Director 3	1050	916
Director 4	1482	906
Director 5	1986	897
Director 6	2553	891
Director 7	3168	887
Director 8	3831	883

Diameter Matters

I have previously discussed the impact of antenna element diameter on both the resonant frequency and bandwidth of antennas. The topic is worth repeating in this discussion about Yagi antennas, since they use different kinds of construction techniques.

Wire Yagis are often used at the lower frequency ranges, where solid rotatable antennas may not be practical. In the upper HF and higher regions up to UHF, Yagis are almost always constructed of tubing supported by a single central boom. The examples I have provided in this chapter include antennas in both camps, wire and tubing.

Yagis at Upper HF

At the low end of the frequency range for rotatable antennas, structures tend to be large and mechanical issues are often paramount. While lower-frequency rotatable Yagis down to 7 MHz are sometimes encountered, the 14-MHz amateur band marks the low end for most Yagi builders. Elements can be 35 feet long and booms even longer on 20 meters. Even with "lightweight" aluminum materials, a 3 or 4 element 20 meter Yagi can be quite a mechanical challenge.

The problem is to have elements that can support themselves, both with static and wind loads, while being secured at a single point in the center. Larger sizes of thick-wall tubing are strong, but they are heavy. Smaller sizes are light, but they bend under the strain of their own weight if they are long. The answer is to use telescoping sections of progressively smaller and lighter tubing as you progress from the center to each end of an element (and boom too). The center section can be heavy to provide strength, without much bending moment, while the sections further from the center can be smaller, with thinner walls since they don't support as much weight. This technique is often referred to as stepped diameter approach. An additional advantage of this technique is that the element lengths can be easily adjusted by loosening the joints and sliding the outer ends in and out.

Table A

14.15 MHz Dipole Lengths for Different Element Diameters

Diameter (in)	Length (in)	Length (ft)	L/D	#/F
0.081	402.5	33.5	4968.9	474.6
1	394.8	32.9	394.8	465.5
1-2 Stepped	405.6	33.8	n/a	478.3
2	390.8	32.6	195.4	460.9
3.75	386.4	32.2	103.0	455.6

Modeling Stepped-Diameter Antennas — the Results

One possible difficulty is how to effectively model tapered elements by computer. The abrupt change in diameter as sections are transitioned is difficult for the NEC engine to handle. Fortunately EZNEC, and other modeling implementations, implement a correction process for these tapering effects.

To give an idea of the differences with and without tapering, Table A shows the modeled resonant lengths of dipole elements resonant at 14.15 MHz for different diameter elements, all mounted 50 feet above typical ground. The 0.81 inch entry represents #12 wire used in my HF example. The 1 to 2 inch tapered entry changes from 2 inches down to a 1 inch diameter, in steps of 0.25 inches. Of particular note is that EZNEC predicts such a tapered element must be longer than either a continuous 1 or 2 inch diameter element.

The entry #/F corresponds to the oft-quoted resonant length of a wire dipole of 468/F, where F is in MHz. That is a reasonable approximation for a wire antenna at lower frequencies but can be misleading when tapered elements of tubing are used.

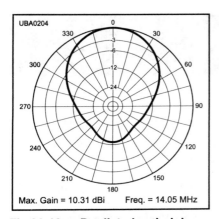

Max. Gain = 10.31 dBi   Freq. = 14.05 MHz

Fig 20-12 — Predicted main-lobe azimuth pattern of the VE7CA tribander on 20 meters. Other bands will be similar.

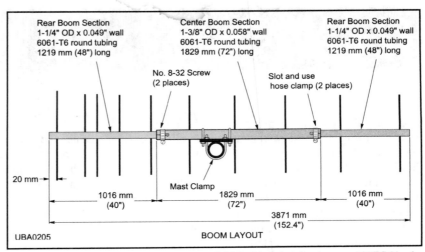

Fig 20-13 — Layout of 10-element 144 MHz high-performance K1FO Yagi.

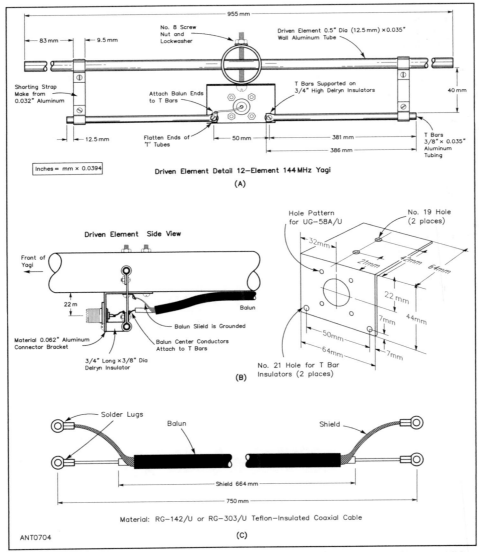

Fig 20-14 — Construction details of 10 element 144 MHz high-performance K1FO Yagi.

Chapter Summary

This chapter discusses building and using Yagi antennas. I presented some of the most commonly encountered methods of matching Yagi driven elements, which tend to exhibit lower impedance than dipoles because of mutual coupling to the other elements. Examples of designs with very different construction methods were presented, including examples of hairpin and T-match arrangements.

Notes

[1] J. Kraus, W8JK, and S. Sturgeon, W8MPH, "The T-Matched Antenna," *QST*, Sep 1940, pp 24-25.

[2] J. Gooch, W9YRV, O. Gardener, W9RWZ, and G. Roberts, "The Hairpin Match," *QST*, Apr 1962, pp 11-14.

[3] M. Hansen, VE7CA, "A Portable 2-Element Triband Yagi" *QST*, Nov 2001, pp 35-37.

[4] M. Hansen, VE7CA, "A Homebrew High Performance HF Transceiver--the HBR-2000" *QST*, Mar 2006, pp 5-9.

[5] R. D. Straw, Editor, *The ARRL Antenna Book,* 21st Edition. Available from your ARRL dealer or the ARRL Bookstore, ARRL order no. 9876. Telephone 860-594-0355, or toll-free in the US 888-277-5289; **www.arrl.orgshop/; pubsales@arrl.org**.

Review Questions

20-1. Consider a center-fed dipole with added reflector and director elements to become a unidirectional 3 element Yagi. Why will there likely be a need to change the method that the coax is connected to the dipole?

20-2. Why can't Q-bars be used to match a 25 Ω impedance Yagi driven element to 50 Ω coax?

20-3. Compare a T-match to a gamma match. What are some of the benefits and disadvantages of each?

Chapter 21

Log Periodic Dipole Arrays

The log periodic dipole array is an effective wideband unidirectional antenna.

Contents

Log Periodic Dipole Arrays

A wideband unidirectional antenna design that is frequently encountered from HF through UHF is called the *log periodic dipole array*, or LPDA for short. This antenna is unique in a few respects:
- An LPDA can provide wideband coverage without adjustment over as much as a 10:1 frequency range, although usually designed to cover a narrower range, typically 2:1 or 3:1 (for example, 10 to 30 MHz, or 150 to 450 MHz).
- An LPDA has a relatively constant gain, pattern and match across the entire range.

The antenna was developed as an outgrowth of studies of other periodic structures by Raymond DuHamel and Dwight Isbell of the University of Illinois in 1957 and has been in wide use since.[1]

What's an LPDA All About?

As its name implies, an LPDA is an array of dipoles. These dipoles are arranged in a special way, with an element-to-element spacing that is logarithmic. The configuration is shown in **Fig 21-1**. Note that the dipoles are fed from the front by a transmission line that is reversed between each successive pair of elements. The combination of out-of-phase feed and Yagi like element length tapering result in radiation focused towards the feed point from those dipoles in the *active region*, as shown in Fig 21-1. The active region contains the dipoles resonant or near resonant at the applied frequency. The LPDA can be fed directly with a 300 Ω balanced transmission line, or fed with coax through a wideband 4:1 or 6:1 balun transformer.

Performance Across a Frequency Range

The dipoles in the active region are responsible for the radiation on a given frequency. The dipoles on the left in Fig 21-1 are resonant at higher frequencies and exhibit a progressively higher capacitive reactance as you move from the active region towards the left side of the array. They thus accept little current and do not contribute to the radiation at the chosen frequency. Similarly, elements to the right of the active region are longer than resonant and appear inductive at the operating frequency.

As the frequency increases, the active region moves to the left and operates in a similar manner and with similar performance until the frequency is higher than just below the resonant frequency of the far-left dipole. As the frequency is reduced the active region moves to the right until it reaches a frequency somewhat below the resonant frequency of the far-right hand element. The far-right and far-left dipole resonant frequencies define the range of operating frequencies for the array.

How Big or Small is an LPDA?

From the discussion above, it should be evident that if the array is lopped off at one or both ends, it should still work, but over a narrower frequency range. That is definitely the case. Small segments of a larger LPDA design are sometime used as driven elements for relatively narrow-band Yagi arrays. In this hybrid Yagi/LPDA array, the Yagi directors help the LPDA driver cell provide a better front-to-back pattern, while the LPDA driver cell expands the SWR

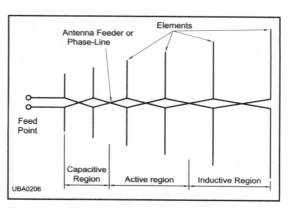

Fig 21-1 — The basic layout of a log-periodic dipole array (LPDA) wideband unidirectional antenna.

bandwidth beyond the usual narrow-band coverage of a Yagi with a single driven element.

In general, the longer the array for a given frequency range, the longer the LPDA active region is and the more elements it must contain. LPDA arrays with relatively large active regions provide more gain and lower SWR ripple across the region. For those who would like to see the design equations, they are straightforward, if somewhat tedious, and are presented in the sidebar. Computer software is available for those who need to perform the calculations often.

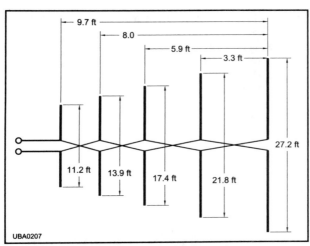

Fig 21-2 — Example of an HF LPDA designed for operation over the 18.1 to 28-MHz amateur bands.

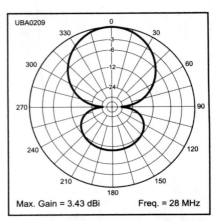

Max. Gain = 3.43 dBi   Freq. = 28 MHz

Fig 21-3 — Typical free space azimuth plot of short LPDA.

So What's Not to Like in an LPDA?

An LPDA can be an excellent choice for some applications, especially those employing wide band signals, such as very wide spread spectrum and especially those employing HF automatic frequency selection (*automatic link establishment* or ALE) based on changes in propagation. In the latter, the radio frequency can shift by a large amount in the midst of communication in response to changes in propagation conditions. This is determined by automatic testing and handshaking between the two ends of the radio link. An LPDA, unlike many HF systems, has no problem following the changes with essentially the same gain and directivity as the frequency shifts — perhaps across the entire HF region.

Perhaps the only downside of an LPDA is that for any given subset of its frequency range, other structures, even narrow-range LPDAs, can be significantly smaller. This is not a bad trade for operations in which the lower frequency end of the range is generally used, with only occasional excursions to the high end, since the majority of the size and mass is being used much of the time. On the other hand, if most of the operations are at the higher end of the range, only a relatively small portion of the mass of the LPDA is used much of the time and other options may be more cost

and size effective.

The LPDA is especially well suited for cases where operation can occur on any part of the spectrum it covers. Some services, such as the Amateur Radio service and the maritime HF service, to name a few, require operation across the HF range, but operate within a finite number of relatively narrow bands within the range. These services can often be supported by multiband Yagi designs that operate on just the multiple narrow bands, but not the whole range. Such systems tend to be smaller and give higher gain than LPDAs covering the entire range.

Other LPDA Configurations

LPDAs are seen in a number of other configurations besides the basic one shown in Fig 21-1. The next most common is one in which the transmission line is not crossed between elements, but the elements are switched from side to side instead. The two sides of the transmission line are generally placed one above the other and the transmission line itself serve as booms for the two sets of elements. Two thick booms provide for a lower-impedance feed than the thin wires of Fig 21-1 and this can provide a good match for 50 or 75 Ω coax — often fed down the inside of one of the booms.

LPDAs are often used as receive antennas for the US television bands. A single LPDA can provide good reception for the VHF channels (including the 88 — 108 MHz FM broadcast band) from 54 to 216 MHz. A single LPDA can cover both the VHF and UHF TV channels (from 470 to 806 MHz). Because of the resulting unneeded coverage over the gap between 216 and 470 MHz, it is more efficient (as discussed in the last section) to have a separate array for the UHF channels. These are gener-

ally combined on the same boom to provide a single feed. The UHF antenna portion could be another LPDA, but often consists of a corner reflector array in front of the VHF LPDA.

Examples of Real LPDA Systems

A version of an LPDA, adapted from an early edition of *The ARRL Antenna Book* is shown in **Fig 21-2**. This was used as the basis for an *EZNEC* model to determine the typical performance of such a system. I designed it to be shorter than optimum (see sidebar), but physically manageable.

A representative free-space azimuth plot is shown in **Fig 21-3** and a 200 Ω SWR plot is shown in **Fig 21-4**. This LPDA is intended to be used with a wideband 4:1 transformer-type balun and then fed by

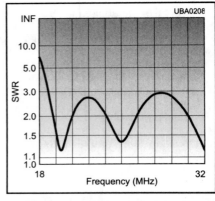

Fig 21-4 — 200 Ω SWR of HF LPDA in Fig 21-2.

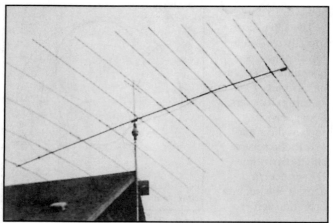

Fig 21-5 — Example of an HF LPDA. This antenna covers 13 to 30 MHz..

coaxial cable going to the station. The performance is less than what can be obtained from a single-band Yagi array on any of its bands, but this LPDA covers all of the 17, 15, 12 and 10-meter amateur bands, as well as everything in-between.

Photos of commercially available LPDA systems for HF and VHF to UHF are shown in **Figs 21-5** and **21-6**.

Fig 21-6 — Another example of a commercial LPDA. This VHF-to-UHF antenna, a Comet/NCG CLP-5130, covers 50 to 1300 MHz, with an advertised forward gain of 10 to 12 dBi. It occupies a 6.7 foot boom. (*Courtesy Creative Design Corp.*)

Designing an LPDA

A full chapter of *The ARRL Antenna Book* is devoted to the design of an LPDA. However, I will here cover the very basics, drawing from that excellent reference.[2]

An LPDA is designed to fit into a truncated triangular shape, as shown in Fig 21-A. Each of the elements just touches the boundary of the triangle, giving the array its characteristic arrow shape. All the element dimensions and spacings are based on the parameters in Fig 21-A. The general design considerations are as follows:

• The smaller the angle α is, the higher the gain and larger the array for any frequency within the range.
• The ratio of highest-to-lowest frequency is roughly proportional to the ratio of longest-to-shortest element, if constructed per the following rules.

The relationship between successive element lengths and spacings can be set to a constant τ (tau) as follows:

$$\tau = \frac{R_{n+1}}{R_n} = \frac{D_{n+1}}{D_n} = \frac{L_{n+1}}{L_n} \qquad \text{(Eq 1)}$$

where R_n and R_{n+1} are the distances from the triangle apex of element n

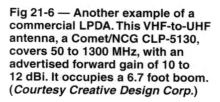

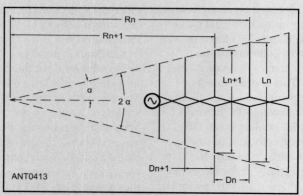

Fig 21-A — Key parameters for the LPDA design equations.

and its nearest inboard neighbor. While I have defined the relationship between adjacent elements and their spacings, I haven't defined the initial spacing. I will thus define another constant as follows:

$$\sigma = \frac{1-\tau}{4\tan\alpha} = \frac{D_n}{2L_n} \qquad \text{(Eq 2)}$$

where D_n is the distance between any two elements of the array and L_n is the length of the longer of the two. For any value of τ, you may determine the optimal value of σ_{opt}, as follows:

$$\sigma_{opt} = 0.243\tau - 0.051 \qquad \text{(Eq 3)}$$

The value for and its corresponding value of yields the highest performance that the LPDA is capable of. The optimum value generally results in a longer boom than is practical, especially in the HF range. Smaller values of σ provide a smaller footprint with corresponding reduction in gain and additional SWR ripple.

Chapter Summary

This chapter discussed log-periodic dipole arrays (LPDAs). These unidirectional antennas are unique in their ability to cover an almost arbitrary range of frequencies with relatively constant gain, directivity and feed-point impedance. They are straightforward to design and construct, although they do tend to be large compared to antennas with similar performance for any single part of their frequency coverage. They are used throughout the HF to mid-UHF range, often as rotatable arrays from the middle of the HF range through the UHF range.

Notes

[1] R. DuHamel and D. Isbell, "Broadband Logarithmically Periodic Antenna Structures," IRE National Convention Record, part 1, pp 119-128, 1957.

[2] R. D. Straw, Editor, *The ARRL Antenna Book*, 21st Edition. Available from your ARRL dealer or the ARRL Bookstore, ARRL order no. 9876. Telephone 860-594-0355, or toll-free in the US 888-277-5289; **www.arrl.org/shop/; pubsales@arrl.org.**

Review Questions

21-1. Consider an LPDA array in comparison to a Yagi. What are the advantages and disadvantages of each?

21-2. Why can't a 4:1 $\lambda/4$ coaxial cable loop balun be effectively used with a wide-range LPDA?

21-3. Describe any other antennas that could fill the place of an LPDA. How do they compare?

Chapter 22

Loop Antennas

This loop antenna, made from plumbing supplies, makes an effective MF receiving antenna.

Contents

This chapter really describes two completely different antennas. The two major divisions are based on the size of the antenna in terms of wavelength. A *small loop* is one whose diameter (or diagonal for a rectangular shaped loop) is much less than a wavelength, while a *large loop* has a diameter approaching or larger than a wavelength.

The Large Loop in a Vertical Plane

I'll start the discussion with the special case of a square loop in a vertical plane, with each side having a length of λ/4.

The Square Quad Loop

Look at **Fig 22-1**, a repeat of Fig 7-1, but with arrows added to indicate the current direction when each source is fed in-phase. This is a broadside array, which has bidirectional gain perpendicular to the page. At a λ/4 spacing, neither the gain nor the end-fire cancellation are optimum, but it would still be a useful performer.

If you bend each dipole towards the other to maintain a λ/4 horizontal section on each dipole, you end up with the configuration shown in **Fig 22-2**. Note that the horizontal sections are still in-phase, while the vertical sections each include the same amplitude of current in opposite phase and thus their vertically polarized radiation is canceled. Note also that the voltages on each side of both gaps are the same, since the two antennas are in phase at corresponding points along their lengths.

Thus, as shown in **Fig 22-3**, with the voltages on each side of the gap the same, you can connect the ends together without changing anything. Having done that, you can remove the upper (or lower, your choice) source and both sides will be fed across the former gap.

It's interesting to compare the performance of such a quad loop with a dipole. If you use my usual HF frequency of 10 MHz and model both at 1 λ above typical ground (to the center of the loop) you get the results shown in **Fig 22-4**, **Fig 22-5** and **Fig 22-6**. Note that the SWR shown in Fig 22-4 has about the same shape and bandwidth as a dipole, but has an impedance at resonance of 123 Ω, rather than the dipole's 70 Ω. This is not much of an issue since you can match it quite nicely to 50 Ω with a λ/4 section of 75-Ω line.

The shapes of the elevation and

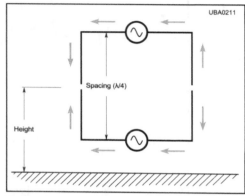

Fig 22-2 – Array of Fig 22-1, with ends bent at 90° to form a square. Note that the currents on the vertical portions are out-of-phase, while the horizontal portions remain in-phase. The two sides of each gap are at the same potential.

azimuth patterns are very similar to those of a dipole at the same average height, except the loop has about a 1 dB advantage in gain at this height. Note the clear tradeoffs between a dipole and a quad loop. A quad loop needs half the spacing between supports. But for comparable low-angle performance, the quad loop needs supports that must be λ/8 higher.

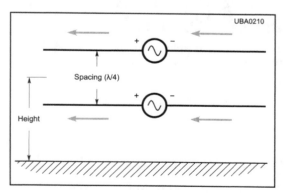

Fig 22-1 – Two horizontal dipoles fed in-phase to form a broadside array. The arrows indicate the direction of current flow.

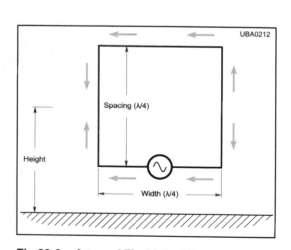

Fig 22-3 – Array of Fig 22-2 with gap closed to close the square. The single source now feeds both "bent dipoles" across the gap.

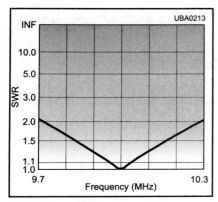

Fig 22-4 — SWR of quad loop referenced to 123 Ω.

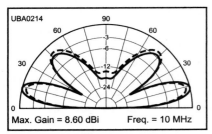

Fig 22-5 — Broadside elevation pattern of quad loop (solid line) compared to dipole (dashed line).

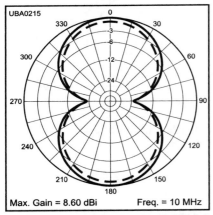

Fig 22-6 — Azimuth pattern of quad loop (solid line) compared to dipole (dashed line) at same average height.

(When both are mounted at the same height, the quad loop's gain at 14° elevation matches the dipole's gain.)

But wait there's more! The loop needs an additional λ/2 of antenna wire, but it needs λ/4 less transmission line. Since transmission line generally costs more than twice as much as wire, that is generally to the plus side for a loop. In addition, there is less sag and tension at the top of the loop, if the wire is supported from the sides.

So the full-wave quad loop provides just a bit of additional gain, and fits into a smaller horizontal space than a dipole. But it needs a minimum of at least λ/3 to λ/2 vertical space to make sense. Otherwise, it acts a lot like a dipole. As I will discuss in the next chapter, a quad loop is frequently used as the basis of multielement arrays.

Quad Loops in Other Shapes

Not surprisingly, most other 1 λ loop configurations have been used, generally with similar results. The two most popular besides the square are shown in **Fig 22-7**. The first, the so-called *diamond* shape, is considered by some to be less vulnerable to ice build up — more important in some geographical areas than in others. Note that the diamond can be suspended from a single support, but requires 1.4 times the horizontal clearance of its square brethren. The single support must be about 0.05 λ higher to have the same effective height, but that doesn't generally matter in practice. If the corners are secured by ropes to the ground, this antenna configuration allows manual azimuth pointing adjustment from the ground.

The triangular, or *delta*, configuration is particularly attractive as a dipole substitute for cases in which the dipole's required λ/2 spacing between supports is not available, although it doesn't have quite the same effective height.

Operating on Other Frequencies

A dipole provides useful radiation on frequencies higher than its resonant frequency, if you can get power to it without loss — such as by feeding it with low-loss open-wire line. The radiation pattern changes dramatically above a full wave in length, moving from being primarily broadside pattern to a multi-lobed one. Such a pattern can sometimes be beneficial, sometimes not. A quad loop is not quite as frequency friendly, as shown in **Fig 22-8**. Note that the result is an elevation pattern pointing mainly upwards. Fig 22-8 shows the endfire view, rather than the broadside direction. Loops are best used on or near their fundamental design frequency, at least until you lay them on their sides, as will be discussed in the next section.

Multiband operation is often accomplished by switching between different loops, one for each desired band. Higher frequency loops are often nested within lower frequency loops. The non-resonant loops usually have sufficiently high impedance that they can be driven together, with the energy going where it is supposed to go. But you must make sure no odd harmonics are involved since those will take power and radiate in strange directions.

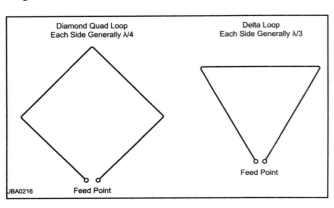

Fig 22-7 — Loops in the diamond and delta configurations. Each is generally 1 λ in circumference.

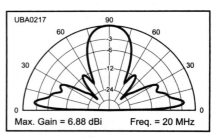

Fig 22-8 — Elevation plot of 10 MHz quad loop operated on 20 MHz.

The Large Loop in a Horizontal Plane

A full wave, or larger loop laid on its side, parallel to the ground, is a useful antenna. This is frequently encountered in amateur circles, and is often referred to as a *Loop Skywire*, a name popularized by a Nov 1985 article in *QST*. The usual configuration is a square or diamond shaped loop about 277 feet in circumference, fed at one corner. This provides a full wave circumference in the 80 meter band and at multiples of 1 λ on many of the higher amateur HF bands. The layout is shown in **Fig 22-9**.

At typical heights of around 50 feet, which is less than λ/4 on 80 meters, the radiation is largely upwards, as shown in **Fig 22-10**. This is well suited for reliable *near vertical incidence skywave* (NVIS) operation out to around 1000 miles or so. The Loop Skywire has useful radiation at lower angles as well. **Fig 22-11** shows the azimuth pattern at an elevation angle of 25°, an angle useful for long-haul operation, with a gain of around that of an isotropic radiator along the diagonal from the corner to the feed point. There is also a notable lack of deep nulls, compared to say a dipole, providing reasonable coverage around the compass.

The feed impedance is similar to the loop in the vertical plane, and for a single band it can be fed through a matching section with coax, as described earlier. The usefulness of this orientation becomes more evident as you move to higher frequency bands, as I shall discuss further below. So it is often fed with low-loss line such as 400 Ω window line. The SWR plot across 80 meters at 400 Ω is shown in **Fig 22-12**.

Operating on Other Frequencies

The Skywire Loop works well at harmonics. The radiation that would otherwise go skyward when a loop is mounted vertically will now leave towards the horizon, providing long haul coverage on the traditional long distance bands. The elevation pattern on 14 MHz is shown in **Fig 22-13**. The azimuth pattern at this frequency is interesting. The pattern at 20° elevation is shown in **Fig 22-14**, which shows four pronounced lobes, each rivaling a small Yagi in forward gain.

The SWR at its second harmonic, 7 MHz, to the top of the HF region is shown in **Fig 22-15**. Note that it is reasonable throughout the range for use with low-loss window line. There are dips of SWR at the harmonics, becoming progressively lower on 20, 15 and 10 meters. This illustrates the flexibility of this antenna in that, much like the dipole operated on higher frequencies with low-loss line, this loop is also useful, with more gain and interesting patterns that are different on each band.

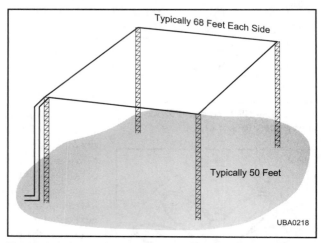

Fig 22-9 — Loop Skywire antenna configuration.

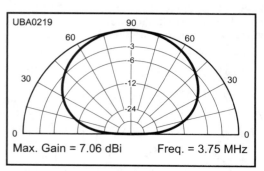

Fig 22-10 — Elevation pattern of Loop Skywire on 80 meters.

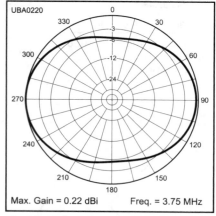

Fig 22-11 — Azimuth pattern of Loop Skywire at 25° elevation on 80 meters.

A Loop Variation — the Terminated Rhombic

The square loop has lobes from each of the wires, and these combine in different directions to form complex lobe structures. These lobes vary as a function of azimuth angle on the higher frequencies. It is possible to adjust the angles of the loop to concentrate the lobes along the major axis of the loop. Tables are available in handbooks to provide the optimum angles as a function of leg length and desired elevation angle. The results are shown in **Fig 22-16** for the loop operated on 28.3 MHz, at which each of the four sides (or *legs*, in rhombic talk) is now 2 λ long. Here, I closed down the angle at the feed from 45° on each side of axis (90° total for a square), to 35° for a 2 λ leg length.

The desired lobe to the right in Fig 22-16 is the result of current running from the source towards the far end. The lobe to the rear results from current reflected from the far end returning towards the source. You can reduce the rearward lobe by terminating the end of the antenna with a resistance that equals the characteristic impedance of the antenna system.

In effect, this antenna looks like a long transmission line. Around 600 Ω is usually optimal, and with such a termination you get the pattern shown in **Fig 22-17**.

The terminated rhombic makes an excellent HF, or even VHF, directional antenna that can be built at low cost. It does take up a lot of space compared to a Yagi, however, and does have a narrow beamwidth, so is best for fixed point-to-point links. If you adjust the dimensions to compromise lengths and angles, it can be effectively used for up to a four-to-one frequency range.

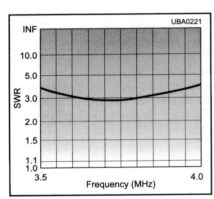

Fig 22-12 — SWR of Loop Skywire on 80 meters for 400 Ω transmission line.

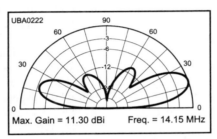

Fig 22-13 — Elevation pattern of Loop Skywire on 20 meters.

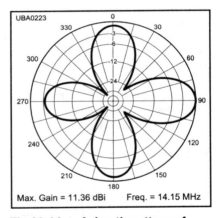

Max. Gain = 11.36 dBi   Freq. = 14.15 MHz

Fig 22-14 — Azimuth pattern of Loop Skywire on 20 meters.

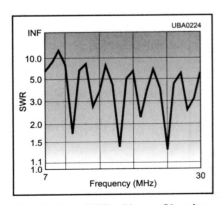

Fig 22-15 — SWR of Loop Skywire from 7 to 30 MHz for 400 Ω line.

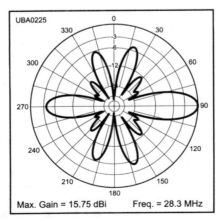

Max. Gain = 15.75 dBi   Freq. = 28.3 MHz

Fig 22-16 — Azimuth pattern of horizontal loop adjusted for optimum 10 meter on-axis pattern.

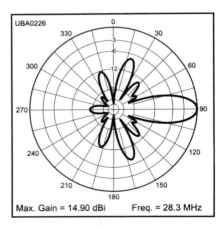

Max. Gain = 14.90 dBi   Freq. = 28.3 MHz

Fig 22-17 — Azimuth pattern of loop modified to be 10 meter terminated rhombic.

The Small Loop Antenna

Loops that are quite small compared to a wavelength (typically less than 0.1 λ in circumference) operate in a very different manner from large loops. Small transmitting loops are sometimes encountered; however, they suffer from having a very low radiation resistance, requiring extraordinary measures to avoid excessive losses. They also have very narrow operating bandwidths — typically in the tens of kHz in the upper HF range. Still, some are commercially available and some have been home constructed with satisfactory results, considering their size.

The most frequent application of small loops is for reception, particularly in the LF to low HF region. A small receiving loop also suffers from the possibility of low-efficiency due to losses, but this is not generally much of a problem for receive loops in this frequency range. The reason is simple: at LF through MF and into the lower HF region (80 meters, for example), received signal-to-noise ratio is limited by external noise picked up on the antenna along with the signal so any loss reduces both noise and signal.

If you're lucky enough to have a 1940s or somewhat earlier AM broadcast radio in your basement, attic or perhaps still in operation in your living room, a look at its back will likely reveal an MF receiving loop. The typical radio of the period had enough sensitivity that it could receive local stations with a multiturn loop, serving as the radio's input tuned circuit as well as a receiving antenna.

A view of this type of antenna is shown in **Fig 22-18**. Note that the multiple turns, as with an electromagnet,

couple to a propagating magnetic field coming from the edge of the loop, the opposite of the orientation of a large, full-wave loop. Note also that the horizontal orientation of the magnetic field implies that a vertically polarized waveform will provide maximum response.

Well, this is just what you want from a receiving loop at the lower frequencies. The vertical polarization is compatible with the ground wave signals arriving from broadcast stations, with their vertical transmitting arrays. The directivity does have deep nulls at right angles to the plane of the loop. Some early broadcast sets had the capability to move the antenna angle so that the whole radio wouldn't need to be reoriented to pick up stations.

After WW II, ferrites became popular as a magnetic core usable well into the radio spectrum. The earlier air-core loops were replaced by compact ferrite core antennas called *loopsticks*. These could use less wire for the same inductance and fit better in the more compact radios of the period. Such antennas were also used in marine-radio direction finders, used to determine the bearings to MF beacons that were located throughout bays and harbors. These were used until the advent of more recent LORAN C and then GPS navigation systems.

As I will discuss in the next chapter, receiving loops can be used to advantage as directional antennas for reception in the MF and low HF bands, discriminating against noise and interference by taking advantage of their directional properties.

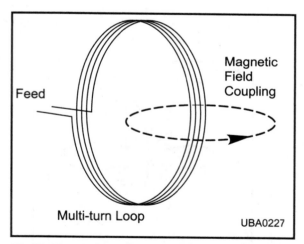

Fig 22-18 — Illustration of a small receiving loop antenna, with its response to an incoming magnetic field.

Chapter Summary

Loop antennas can take a number of forms and each has a special place in radio. Large loops are efficient HF radiators that can be used by themselves or as part of arrays. Vertically oriented loops give dipoles a run for their money, and horizontal ones can provide considerable flexibility from a single antenna.

While a small loop can be force fit into transmitting applications, it shines the most if used for MF receiving. In this application, its advantages can be taken advantage of without excessive concerns about its loss limitations.

Review Questions

22-1. Consider a vertically oriented loop compared to a horizontal dipole. What are the relative advantages and disadvantages of each in the same space?

22-2. What are the benefits of large horizontal-plane loops compared to vertical-plane loops on harmonics of their primary resonant frequencies?

22-3. Compare the benefits of small loops for receiving with the use of such loops as transmitting antennas.

Chapter 23

Loop Antennas You Can Build

This loop antenna transmits and receives on 40 meters or can tilt down for transport.

Contents

Loop Antennas You Can Build

The loop antennas described in the last chapter are not only good performers, especially considering their low cost, but they are generally easy to build and get operating. They tend to be dimensionally forgiving and generally non-critical as to construction method compared to some other antenna types. The designs presented here can be scaled in frequency to work other bands without undue difficulty.

A Quad Loop for HF

The quad loop is an effective antenna, especially for HF operation. The antenna is inexpensive to make, has less *wingspan* than a λ/2 dipole and avoids the center droop and mechanical loading of an end-supported dipole with heavy coax feed at its center. Depending on the spacing between supports and their height above ground, it can be square, rectangular or delta shaped.

The key dimension is the total length of the loop wire, although, as noted previously, the length depends to a certain extent on whether the loop is square or triangular in shape. For a square loop with top 88 feet above typical ground, the length in feet is around 1030/F (MHz) at 3.8 MHz.

A triangular loop will require about 1016/F (MHz) at 3.8 MHz. These guidelines were established at 3.8 MHz using *EZNEC* models. The modeled lengths at some other frequencies are shown in **Table 23-1**. Height above ground, nearby objects and ground conditions will also make a difference in resonant frequency, so always start a bit long and trim a little at a time to move the resonant frequency to the part of the band you want. Construction suggestions are provided in **Fig 23-1**.

There is a slight performance difference between the square quad loop and the delta loop configurations, mainly due to the effective height above ground. The square loop will have an effective height

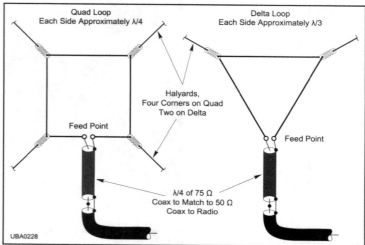

Fig 23-1 – Loops in the "quad" and "delta" configurations. Each is generally 1 λ in circumference.

equal to that of its center, half way between the two horizontal sections, as you might expect. The delta configuration, with the apex at the bottom will have an effective height somewhat higher. By comparison, a horizontal dipole will have its effective height at or slightly above the height of its center, depending on the amount of sag.

While the full-wave loop has a gain of almost 1 dB over a dipole in free space, closer to the earth and constrained by the same support height, the dipole actually has a bit of an edge at low angles. This is shown in **Fig 23-2** and in

Table 23-1

Comparison of Dipole and Loop Dimensions (feet) on Some Amateur Bands. Top Height for all is 88 Feet.

Band (Meters/MHz)	80/3.75		40/7.15		20/14.15		10/28.3	
Antenna	Width	Height	Width	Height	Width	Height	Width	Height
Dipole	128	Nil	66.6	Nil	33.6	Nil	16.8	Nil
Delta Loop	91.4	79.2	48.7	42.2	24.6	21.3	12.4	10.7
Square Loop	67.6	67.6	36.2	36.2	18.3	18.3	9.2	9.2

Table 23-2

Comparison of Dipole and Loop Parameters and Performance.

3.8 MHz Antenna	Width (feet)	Elevation Angle	Peak Gain (dBi)	Gain at 42° Elevation (dBi)	Impedance (Ω)
Dipole	126.4	42°	6.29	6.29	90
Delta Loop	90.4	47°	5.72	5.64	130
Square Loop	66.8	50°	5.31	5.16	120

Table 23-2. The difference shows up mainly in the elevation angle of the main beam, with lower angles needed for longer distances. As shown, the differences are about 1 dB at most, and thus the physical and mechanical considerations are often the decision makers.

Besides height, the other aspect of any trade-off is the space required between horizontal supports. Table 23-2 provides the dimensions and performance figures for a square quad loop, an equilateral delta quad loop and a horizontal dipole at 3.8 MHz, all at a top height of 88 feet, corresponding to the height of the delta apex being 10 feet above the ground.

Dimensions for some amateur bands are provided in Table 23-1 for the bare #12 wire used in the modeling. In all cases, the antennas are suspended from 88 foot high supports. If insulated "house wire" is used, reduce the dimensions by about 2%.

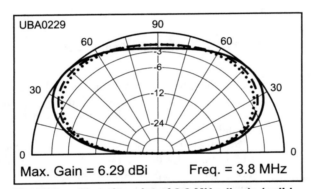

Fig 23-2 — Elevation plot of 3.8 MHz dipole (solid line), delta (dashed line) and quad loop (dotted line) at same top height.

Two Element Cubical Quad for 20 Meters

One of the most popular HF loop antenna systems is the two element *cubical quad*. This is essentially a two element Yagi using loop elements. While more than two elements are sometimes found, the two element version is the most popular. It is often said that a two element quad is similar in performance to a three element conventional Yagi, due to the free space gain of a full-wave loop over a dipole. This difference is maintained with more elements but a point of diminishing returns can be reached in terms of array size.

Still, at the two element point, the quad is quite attractive. It has a width about half that of a conventional Yagi, and its boom length is about half of a three element Yagi with comparable gain. Its elements can be constructed of wire supported by an inexpensive structure and its higher feed impedance is much easier to directly feed than that of a Yagi. There's a lot to like in this picture!

Building a Quad

The only tricky aspect of a quad array is that the structural elements holding the corners of the loops can't be solid metal. A typical square cubical quad is shown in **Fig 23-3**, along with construction suggestions. Key

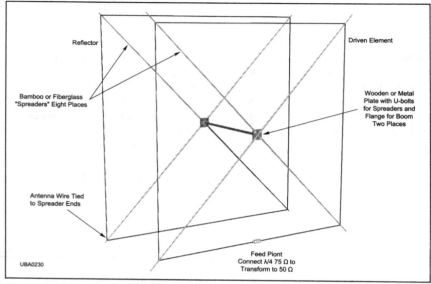

Fig 23-3 — Cubical quad construction.

dimensions are shown in **Table 23-3** for a quad made of #12 wire with the antenna's center 50 feet above the ground.

Note that I have selected a 10 foot boom for this example. A 13 foot boom would provide a bit more gain on 20 meters; however, 10 feet is an easier length to obtain and works better for higher bands if you wish to add them later. The LENGTH × FREQUENCY entry in Table 23-3 is there in case you wish to move the center frequency. Just divide the value by the desired frequency (in the same band) and you will have the new length in feet. These numbers are somewhat different than the oft-published ones, and reflect this band at this height and element spacing.

The spreaders can be made from bamboo poles, which are sometimes available

free from carpet dealers. Alternately, there are fiberglass poles made for fishing that may be suitable. If bamboo is used, it should be given a few coats of exterior varnish to increase its weather resistance.

The boom should be made from 1.5 or 2 inch thick walled aluminum tubing, depending on expected wind load and weight of the completed element assembly. Modeling showed no discernible difference with or without an aluminum boom and a 10 foot aluminum mast at its center, so materials for these functions do not impact electrical performance in the same way that they do for the element spreaders.

So How's it Play?

The modeled performance is shown in **Fig 23-4** and **Fig 23-5** for a height of 50 feet above typical ground. Note the nice azimuth pattern and the sharp front-to-back ratio. The forward gain of more than 12 dBi is more than 4 dB higher than a simple quad loop without a reflector at the same height. The three element Yagi discussed in Chapter 20, adjusted at 14.15 MHz at a height of 50 feet has a forward gain

Table 23-3

Key Dimensions of 20-Meter, two element Cubical Quad for 14.15 MHz.

Parameter	Length
Total Driven Element	70 feet, 2 inches
Length × Frequency	993
Driven Element Spreader	12 feet, 5 inches
Total Reflector	73 feet, 7 inches
Length × Frequency	1041
Reflector Spreader	14 feet, 7 inches
Boom	10 feet
Matching Section (75 Ω Poly, 0.66)	11 feet, 6 inches
Matching Section (75 Ω Foam, 0.8)	13 feet, 11 inches

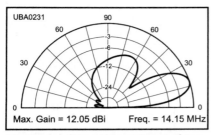

Fig 23-4 — Elevation pattern of two element 20-meter cubical quad.

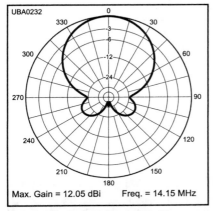

Fig 23-5— Main-lobe azimuth pattern of two element 20 meter cubical quad.

of 12.3 dBi, so the quad is definitely a contender in a smaller, perhaps less expensive and more compact package.

The 100 Ω SWR curve is shown in **Fig 23-6**. Note how flat it is across the band. It could be fed directly with 75 Ω coaxial cable, or better, with 93 Ω coax, if available. The usual arrangement, however, is to transform the impedance through a λ/4 section of 75 Ω cable. The result is an almost perfect match to 50 Ω coax at mid band. Table 23-3 shows the required lengths of solid and foamed polyethylene dielectric coax using the typical values shown. If you have manufacturer's data available, adjust for published relative dielectric constant.

Adding Additional Bands to the Two Element Quad

It is possible to use the same spreaders and quad infrastructure to support another quad for additional bands. For the case of a 20 meter quad, you could choose any of the four higher frequency HF amateur bands and install the quad wires symmetrically inside the two 20

meter loops. Note that a 10 foot boom length is 0.22 λ at 21.2 MHz, a good spacing for a two element beam.

The resulting dimensions are shown in **Table 23-4**. Note that while the unmodified 20 meter quad still operates on 20 meters following the addition of the second quad, although its optimum gain and front-to-back ratio has moved down about 150 kHz. For this reason, I have shown a new total length for the 20 meter reflector. The resonant frequency was not significantly changed so I did not change the driven element's dimensions.

The 15 meter quad's performance is shown in **Fig 23-7**, **Fig 23-8** and **Fig 23-9**. The gain is a bit higher due to the wider spacing. The 20 meter performance is unchanged, following the adjustment of reflector length.

Feeding the Two Band Quad

The best way to feed the two-band quad is by having a λ/4 transformer of 75 Ω coax from each feed point and then either separate 50 Ω transmission lines back to the station, or a single 50 Ω line with a remotely controlled relay. Some have simply connected the feed points together and used a single line; however, my modeling indicates that while the performance does not seriously suffer, there will be high SWR on at least one band. This will result in

increased feed-line loss, with a net decrease in system gain, as well as a more complicated tuning network at the transmitter end.

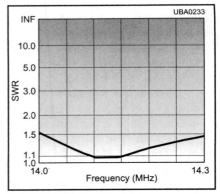

Fig 23-6 — 100-Ω SWR of two element 20 meter cubical quad.

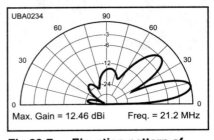

Fig 23-7 — Elevation pattern of two element 15 meter cubical quad using the same boom length as the 20 meter quad.

Table 23-4

Key Dimensions of Interlaced 20 and 15 meter, two element Cubical Quad for 21.2 and 14.15 MHz.

Parameter	Length
New 20 Meter Reflector	72 feet 10 inches
Total 15 Meter Driven Element	46 feet, 7 inches
Length × Frequency	987
Driven Element Spreader	12 feet, 5 inches
Total 15 Meter Reflector	50 feet, 5 inches
Length × Frequency	1068
Reflector Spreader	14 feet, 7 inches
Boom	10 feet
Matching Section (75 Ω Poly, 0.66)	11 feet, 6 inches
Matching Section (75 Ω Foam, 0.8)	13 feet, 11 inches

Adding Even More Bands

Quad builders have been known to build quads with wires for as many as five bands on a single set of spreaders. There is no reason that this could not be done; however, I have a few caveats:

- Expect to have to deal with a fair amount of interaction and subsequent retuning. You may want to provide a mechanism for easy adjustment of element length.
- The element spacing at one or the other ends of the range is likely to suffer, resulting in performance far from optimum on at least one band. A way to avoid this issue is to mount the spreaders in such a way that the spacing between elements increases as you get closer to the outer end of the spreader.

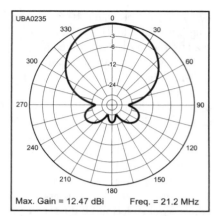

Fig 23-8— Main lobe azimuth pattern of two element 15 meter cubical quad.

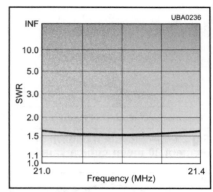

Fig 23-9 — 100-Ω SWR of two element 15 meter cubical quad.

The Loop Skywire

The Loop Skywire introduced in the last chapter is another popular antenna among Amateur Radio operators. The basic configuration is shown again in **Fig 23-10**. You use a 1 λ loop on the lowest frequency of operation. Many people use it on other bands as well. The impedance varies by band as shown in **Table 23-5** for a 272 foot overall length, square loop at a height of 50 feet. Note that unless the 80 meter resonance is at the very bottom of the band, the harmonic resonances are not in quite the right place. While some people feed the Loop Skywire directly with 75 Ω coax, the SWR is quite high on the higher frequency bands, partly masked at the bottom end of the coax due to the loss caused by the high SWR. A better solution is to use low loss open

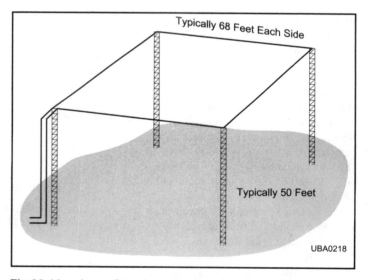

Fig 23-10 — Loop Skywire antenna configuration, 272 feet overall length.

Table 23-5

Impedance of 277-Foot Long Loop Skywire on Different Bands.

Frequency (MHz)	Resonant Impedance (Ω)
3.65	137
7.35	100
10.9	240
14.45	235
21.5	240
28.6	234

wire feed line and then to use an adjustable matching device (antenna tuner) at the bottom to transition to 50 Ω coax.

Performance charts for the antenna are shown in Chapter 22. I have heard from readers who had limited space but who were successful with all sorts of loop shapes, some at much lower heights. Perhaps this is based on comparisons with other, even more unsatisfactory antennas, but the Loop Skywire seems to be flexible even in such problematic applications!

Low Noise Receiving Loop

From the middle HF frequencies (from 14 MHz upwards), the usual challenge for a receiver is to capture the maximum signal it can, since the received signal-to-noise ratio (SNR) is generally limited by noise within the receiver at those frequencies. Every additional 0.1 μV helps!

Below about 14 MHz, depending largely on the time of year, the location and receiver specifications, much more noise is received by the antenna than is generated within the receiver itself. Much of this noise comes from thunderstorm static crashes, plus electrical noise from arcing powerlines and other such manmade sources. Under such circumstances, increasing the amount of signal captured generally increases the received noise, also, with no improvement in the signal-to-noise ratio (SNR).

This situation allows us to design antennas specifically for improved SNR, rather than for maximum signal. A small receiving loop is one antenna that can be used to improve SNR. The improvement is a result of a number of factors. First, a small loop is quite directional along the plane of the wires in the loop, so it can be rotated to find the azimuth with maximum SNR, often by placing specific noise source arrival directions in its deep nulls. In addition, it is responsive to vertically polarized signals arriving from low angles, rejecting horizontally polarized signals that may come from nearby (or even distant) noise sources.

There are a number of published designs; however, the one that follows is from an article by Richard Marris, G2BZQ, in *QST*, Oct 1985,

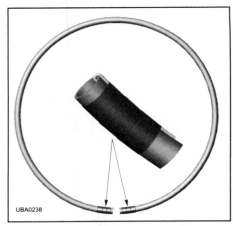

Fig 23-12 — Pull L2 (2 mm OD stranded PVC insulated wire) through the tubing first and secure its end with tape while L1 (0.6 mm solid wire with PVC insulation) is being wound. If you have trouble pulling the length of all six turns through, it may be cut and spliced at the gap after pulling it all through.

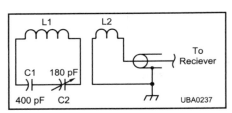

Fig 23-11 — Schematic of the HH160RL. Both L1 and L2 are within the tubing. C2 is a receiving-type variable capacitor. If a capacitor of at least 280 pF, or a common 365-pF receiving type is used, C1 is not needed. L1 is a six-turn loop; L2 a single turn.

pp 45-46. It is straightforward and easy to reproduce. Richard called his the *HH160RL*, for Hula Hoop 160 meter receiving loop. He built his from a discarded Hula Hoop, but you can use any 24 inch length of semi-flexible plastic tubing with an inside diameter of at least ⅝ inches.

The circuit of the receiving loop is shown in **Fig 23-11**. Note that the very narrow bandwidth and low radiation resistance of such a small antenna make direct connection inefficient. It is common, as in this design, to transformer couple to a pick-up loop mounted within the main loop to avoid interfering with loop tuning and efficiency.

Construct the loop as described in **Fig 23-12**, **Fig 23-13** and **Fig 23-14**. A split bamboo or PVC plug that fits into and joins the tubing ends around the wires can be used to hold the ends together. Slide the plug into each end with the loop wires in the middle until there is a gap of about ¼ inch. Use epoxy (or for a PVC joint, PVC cement) to hold it together. Mount the loop in the chassis box, wire per the schematic and try it out. Note the directional pattern in **Fig 23-15**, the oppostie of that of a large loop.

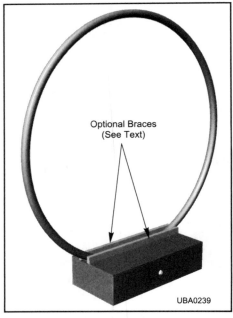

Fig 23-13 — The loop is installed in a wooden or metal chassis box containing the variable capacitor with wood braces as shown.

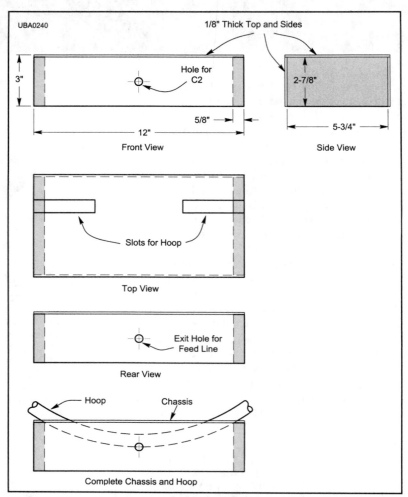

Fig 23-14 — Detailed views of the chassis construction.

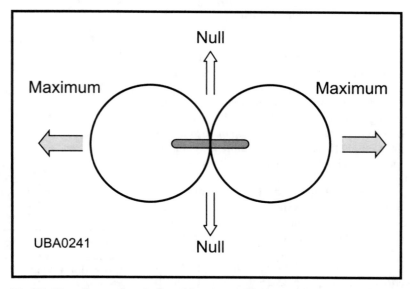

Fig 23-15 — Reception (azimuth) pattern of the receiving loop. It is turned to make best use of its sharp null to minimize noise.

Chapter Summary

Large loop antennas can be effectively used for both transmitting and receiving and offer some physical advantages over other types of structures. There are a number of configurations and each has a special place in antenna technology.

Small loops can be made into very narrow-band transmitting antennas, but really shine as directional MF receiving antennas. They can improve received SNR while taking little space and costing little.

Review Questions

23-1. Which antenna, strung from the same halyards, has the most of its radiation at lower angles — a dipole, square loop or delta loop with apex at the bottom? Which is next? What accounts for the difference?

22-2. Why is low-loss line recommended for a Loop Skywire on the harmonics of its 1λ resonant frequency?

22-3. Why would an indoor low-noise receiving loop not likely to be of significant benefit at VHF?

Chapter 24

Antennas for Microwave Applications

Daryl, KG4PRR, installs a microwave reflector array.

Contents

Antennas for Microwave Applications

The antennas I've discussed so far have been what I would call traditional, in that they are connected to transmitters via transmission lines. At the higher UHF into microwave frequencies, the losses associated with transmission lines can be avoided by using other technology for RF power delivery.

Waveguide

One such system is called *waveguide*, a replacement for coaxial cable in the upper UHF to SHF range for many applications. A waveguide is a completely different kind of transmission medium from traditional coaxial or other transmission lines. Transmission lines have loss that goes up logarithmically with frequency. For example, a very satisfactory 50 Ω coax cable for HF, such as RG-213, will have a matched loss for 100 feet at 1 MHz of about 0.2 dB, about 0.6 dB at 10 MHz, 2 dB at 100 MHz, 8 dB at 1 GHz and about 30 dB at 10 GHz.

The reason for the increase of loss with frequency is simple: there are two primary loss mechanisms. The first is the resistance of the conductors, primarily the inner conductor because of its smaller size. The resistance goes up with frequency due to *skin effect*.

Losses within the dielectric between the conductors in a coax cable generally also increase with frequency, for any kind of solid material.

Since power at microwave frequencies is usually expensive to generate, coaxial cable is generally found only in very short interassembly cables or for noncritical functions.

A waveguide is a metallic duct structure that accepts a propagating wave and supports transmission through the length of the duct with minimal attenuation. In order to do this, waveguide must be of a size comparable to a wavelength. The inside surface of the duct is a good conductor — often silver plated — that establishes boundary conditions allowing fields that support propagation down the guide. The fields are excited inside one end of the waveguide by small antennas that form a transition between circuit elements, including transmission lines, on the outside and propagating fields on the inside of the waveguide.

One way to think about waveguide is as an extension of open-wire transmission line. **Fig 24-1** shows an open-wire transmission line with a single λ/4 shorted stub in parallel with the line. The shorted stub has an infinite impedance at its open end and thus can be bridged across the line without changing the flow of energy down the line. You could continue adding such stubs on both sides of the line until they become a continuous tube with a rectangular cross section. The result is a waveguide that will transport RF at a frequency corresponding to its width of 2 × λ /4, or λ /2. Signals with a frequency much below this *cutoff* frequency will not be propagated, supporting the size-to-frequency relationship of the waveguide.

Although I've seen waveguide used in high budget military radars at frequencies around 450 MHz, most are used at frequencies above 2 GHz, allowing reasonably sized structures. By 10 GHz, the size of waveguide approaches that of a large coaxial cable. Waveguide is most often found in rigid sections, straight or angled, and interconnected by flanges. Some semi flexible waveguide is available, making it almost as easy-to-handle as semirigid coax.

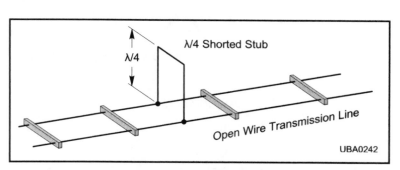

Fig 24-1 — Open-wire transmission line with λ/4 shorted stub.

Horn Antennas

A number of popular microwave antennas follow directly from the use of waveguide, although they are not restricted to waveguide use. The first I shall discuss is the *horn* antenna. The horn antenna, in its simplest form, is just a waveguide opened up on the end to let a wave propagating in the waveguide escape and become a wave propagating in space.

The horn antenna is essentially a flared out section of waveguide, in one or both dimensions. By gradually expanding the dimensions, the field can be smoothly transitioned into a plane wave with a larger aperture, providing defined directivity in either or both axis. For usual waveguide modes, the electric-field vector is across the narrow direction, roughly correlating to the voltage across a transmission line model, as seen in Fig 24-1 and **Fig 24-2**, with the magnetic-field vector in the direction aligned to the larger dimension of the waveguide, and the electric field vector perpendicular and across the narrow dimension, as shown in **Fig 24-3**. Thus, either vertical or horizontal polarization can be selected, merely by orienting the waveguide section before the transition to the horn.

Rectangula cross-section waveguides transition naturally to rectangular or pyramidal horns, while circular or elliptical waveguides more naturally feed into cylindrical or oval-shaped horns. While waveguide feed of horns is a natural fit, horns can be transitioned from coaxial cable through a very short section of closed waveguide with a coax-to-waveguide coupling section.

Horn Dimensions

The key dimensions of a rectangular horn are the length (L) and aperture size (A_E and A_H). These are shown in **Fig 24-3**. For a circular horn, the corresponding aperture dimension is the common diameter (D). The dimensions are related in that for

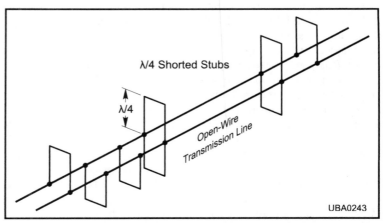

Fig 24-2 — Open-wire transmission line with multiple λ/4 shorted stubs.

a given aperture (and hence beam-width and gain), there is an optimum horn length that provides uniform illumination to allow the effective use of that aperture. The relationships are defined in **Table 24-1**.[1]

The resulting gain of the horn follows from the directivity and is approximately given by:

$$\text{Gain (dBi)} = 10\log_{10}(7.5 \times A_E A_H)$$

(Eq 1)

Horn Applications

The horn can be, and often is, used directly as an antenna. Perhaps more commonly, however, it can be placed at the focus of a parabolic-dish reflector and used to drive a higher-gain system. Through the adjustment of the above parameters, the horn

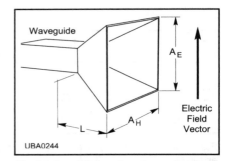

Fig 24-3 — View of rectangular horn antenna terminating a waveguide.

can provide uniform illumination of the dish with minimal waste of dish aperture, or of power being wasted by illuminating area outside the dish. In addition, the dish offers almost infinite front-to-back ratio, reducing extraneous responses in undesired directions.

Table 24-1

Beamwidth of Horn Antenna as a Function of Key Dimensions (λ)

Aperture	Beam Width (degrees)	
	Between First Nulls	Between 3-dB Points
Length for uniform illumination	115/L	51/L
Rectangular horn E-plane	115/L	56/A_E
Rectangular horn H-plane	172/L	67/A_H
Circular horn	140/L	58/D

Slot Antennas

The *slot antenna* is an interesting configuration that truly requires "thinking outside the box"! **Fig 24-4** shows a pair of λ/4 shorted stubs, used previously to describe a waveguide. In this example, however, they are fed at the center. This won't make a very good antenna because the currents on opposite sides are out-of-phase, equal and so close that radiation is essentially cancelled. If you take this case and turn it inside out by cutting a λ/4 slot in a sheet of metal, the currents will flow as shown in **Fig 24-5**, migrating from the immediate area of the slot. The result is an efficient radiating structure that can be used as an antenna. The radiation will be bidirectional towards both sides of the sheet and will be vertically polarized if oriented as shown. The impedance at the center is about 500 Ω. By feeding it about λ/20 from one end of the slot, a good match to 50 Ω coaxial cable can be obtained.

A slot can also be fed directly by a waveguide. By merely attaching a flat plate, typically a minimum of 3λ/4 wide and λ/2 high with a slot matching the waveguide opening, you have a form of unidirectional slot. It will radiate fairly uniformly across the hemisphere on the open side. Unfortunately, my antenna modeling tools only work with wire antennas, so I don't have the capability to provide the usual modeled results for these antennas.

A unidirectional slot antenna can also be made by putting an empty box behind the opening with a depth of about λ/4. This could also be considered a λ/4 shorted-waveguide stub and can serve as a very nice reflector. The applications for metal vehicle bodies should be apparent, especially since the slot need not be air, but could be some kind of dielectric structure to avoid turbulence.

To obtain additional directivity, multiple slots can be spaced and oriented just as the linear phased arrays discussed for HF. A multiple-aperture slot array can also be constructed by putting multiple slots in the appropriate places on the surface of a waveguide.

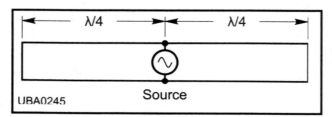

Fig 24-4— Two λ/4 shorted stubs in parallel, fed as an antenna.

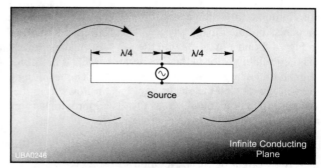

Fig 24-5— Slot-antenna configuration.

The Patch Antenna

An interesting structure is called the *patch antenna*. It is constructed from a simple rectangle of metal, λ/2 on a side, suspended over a metal ground plane. This is shown in **Fig 24-6**. The coaxial feed is connected inward from the center of an edge at a point that matches the coaxial-cable impedance, with the shield just below on the ground plate. Current flows across the patch radiating in a direction perpendicular to the plate. The pattern and gain are similar to a pair of in-phase dipoles, spaced λ/2, positioned in front of a reflector. The resulting gain is about 8 dBi with a 3 dB beamwidth of about 60°. If the ground plane were infinite in extent, the front-to-back ratio would also be infinite. With a small ground plane, somewhat larger than the patch, the F/B will be about as with a Yagi.

The patch is a versatile antenna that can be found in a number of applications. Its ease of construction and predictable response make it popular as a feed antenna for home-made dish antenna systems. It can also be easily fabricated on double-sided printed circuit board, although the dielectric material will result in some changes in performance and resonance. If fed directly on the edge, strip-line transmission-line sections can be easily etched at the same time to result in a match to whatever the system designer needs as a load.

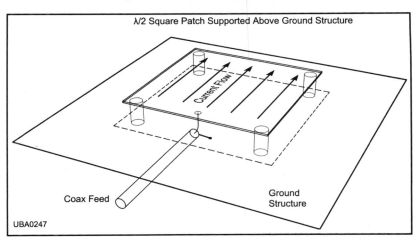

Fig 24-6 — Construction of a patch antenna over a finite ground plate. The radiation is perpendicular to patch.

Chapter Summary

At frequency ranges with small wavelengths, different antenna structures are possible than at the lower frequencies where I've spent most time in this book. Keep in mind, though, there is nothing directly related to frequency associated with these structures and they could be built for lower frequencies as well if there were particular requirements for them.

Notes

1,2 J. Kraus, *Antennas*, second edition, McGraw-Hill Book Company, New York, 1988, pp 651-653.

Review Questions

24-1. What is it about microwave antennas that make them different than antennas for lower frequencies?

24-2. Why are long coaxial cable runs not often encountered at microwave frequencies?

24-3. What are the relative advantages of patch and horn antennas as feed systems for dish antennas?

Chapter 25

Vehicle Antennas

There's almost no limit to the antennas you can support from some vehicles.

Contents

Vehicle Antennas

Most of the antennas I have covered so far have been developed assuming that they were mounted some distance from the ground and thus were pretty much alone in their environment. This works well for antennas suspended a distance above ground from halyards or on a structural support such as a tower. There is another category that shares some of the previous characteristics, but that brings special challenges — antennas mounted on vehicles.

In this context, by vehicles I mean land-operating motor vehicles, aircraft and water-borne vessels. Each may need to be equipped with radio communication or radionavigation systems, and each brings its own set of challenges.

Land-Operating Motor Vehicles

We often take for granted the telecommunications capabilities of modern motor vehicles. Many have receive capability at MF (AM broadcast), VHF (FM broadcast) and UHF (satellite-based GPS navigation) as well as UHF two-way communication for cellular telephone, or specialized service programs (On-Star). In addition, most public service and utility vehicles provide dedicated two-way VHF or UHF radiotelephone and often data systems. There are also some two-way mobile HF systems in common use, principally by Amateur Radio operators and military vehicles. It's amazing that among these, only the occasional Amateur Radio operator's vehicle looks a bit like a porcupine or a fishing trawler!

Receive-Only Antennas

MF receive antennas go back in time to just after it became possible to make compact AM broadcast receivers that could fit into car dashboards. The typical antenna is a short vertical monopole, much shorter than λ/4. While a full-size λ/4 monopole would be quite a good performer at 550 kHz, the bottom of the AM broadcast band, it would be about 425 feet tall. This would be a mechanical challenge, not to mention the problems associated with going through the drive-through lane of your bank!

The typical antenna is a rod 3 or 4 feet high. It receives a much smaller amount of the usual broadcast ground-wave signal than a full-sized antenna, but because the SNR is limited by external noise, the signal is strong enough to provide usable reception. An MF automotive receiver generally has additional RF amplification compared to a home radio, to bring the received signal above that generated within the receiver.

VHF-FM broadcast reception is much more straightforward. The FM broadcast band extends from 88 to 108 MHz in the US. At 100 MHz, λ/4 is just 1.5 meters or about 59 inches long, not much longer than the length of the shortened MF monopole. With the ground side connected to the vehicle body at the base of the antenna, it can be easily matched to the input of the receiver. The reception will be fine for line-of-sight (LOS) paths to transmitters, and broadcast transmitters are generally well elevated to account for low receive antennas.

Reception of GPS satellite signals adds a new dimension to the vehicle problem. In the previous cases, I have discussed reception from transmitters sending signals at low angles of arrival — from LOS or ground-wave paths. Whatever we expect from satellites, it won't likely be restricted to low-angle reception.

The elevation pattern of a λ/4 monopole over a near perfect ground,

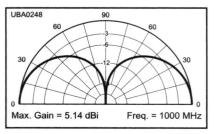

Fig 25-1 — Elevation pattern of 1-GHz λ/4 monopole on vehicle.

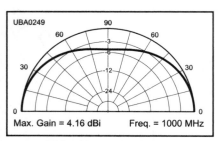

Fig 25-2 — Elevation pattern of 1-GHz λ/4 bent monopole or inverted L on vehicle.

such as a metal vehicle roof at UHF, is shown in **Fig 25-1**. It clearly will be much better at reaching low-angle cellular towers than high-angle satellites for GPS or satellite-radio use. By bending about ⅔ of the vertical portion over to become horizontal, you form an antenna that would be called an *inverted L* in the MF and HF region.

The resulting elevation plot is shown in **Fig 25-2**. Note that while the low-angle response is only down

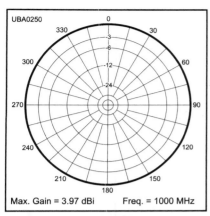

Fig 25-3 — Azimuth pattern of 1-GHz λ/4 inverted L at 15° elevation angle.

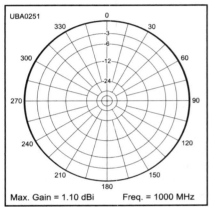

Fig 25-4 — Azimuth pattern of 1-GHz λ/4 inverted L at 60° elevation angle.

about 1 dB, the high-angle response is significantly increased. Lest you wonder about the azimuth coverage, low- and high-angle plots are shown in **Fig 25-3** and **Fig 25-4**. These show that the uniform response is maintained. The SWR of the antenna at 1000 MHz is shown in **Fig 25-5**. It is moderately sharp, but could easily be adjusted for either GPS operation at 1575 MHz or cellular coverage at either 860 or 1900 MHz.

An improvement of this approach is the popular wideband *blade* antenna. This is an adaptation of the slot antenna described in Chapter 24, and will be discussed further in the aircraft section that follows. The blade antenna can cover the entire range and looks a bit like a small dorsal fin on the back of many recent sedans.

Two-Way Radio Antennas for Motor Vehicles

Antennas for effective transmission and reception from motor vehicles divide nicely by the portion of the radio spectrum into which they fall. In the VHF and UHF ranges, the most common antenna is a vertical monopole. It provides the omnidirectional coverage sought by most vehicle systems. Such an antenna is quite efficient, since the size of the typical vehicle body makes for a good ground system under the antenna. Typical lengths range from λ/4 to 5λ/8, with the longer sizes used mainly at higher frequencies where the physical length is still reasonable.

The major limitation of such systems is that they are essentially limited to line-of-sight paths, and those generally provide limited range between land vehicles. Most VHF/UHF systems operate between a base station with an elevated antenna and multiple mobile units, or via remote repeater stations in high terrain that can "see" all mobile units over a wide geographical extent and relay transmissions between them.

HF and even MF two-way mobile systems have been in use for a long time, in spite of the fact that efficient, reasonably sized antennas for that frequency range are hard to come by. Before municipal police departments moved into the VHF range somewhat after WW II, they used MF frequencies just above the broadcast band for communications between headquarters and mobile units.

The challenge with HF mobile operation is that the same λ/4 monopoles that work well at VHF are much bigger at HF. At the top of the HF range, 30 MHz, a λ/4 monopole is about 8 feet long, and it appears often as a stainless steel flexible *whip* antenna operated against the vehicle body as a ground. The US citizen's band

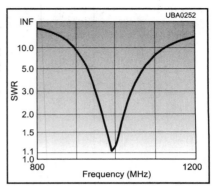

Fig 25-5 — SWR plot of 1-GHz λ/4 inverted L.

(CB) at 27 MHz is just at the edge of practicality with full-size antennas 9 feet long, frequently seen mounted on trucks.

For lower-frequency HF operation, electrically short antennas are usually used. It's necessary to compensate for the shorter length in some manner. The choices are to have a whip antenna with a matching network beneath the vehicle body, or to electrically lengthen the monopole by *loading* it. This term, borrowed from telephone company line-compensating parlance, means substituting an inductor for part of the antenna length to provide a resonant antenna that is physically shorter than λ/4. The *loading* inductance can be mounted at the base or somewhere in the whip itself and is sometimes used in combination with a capacitance at the top. A loaded mobile antenna is shown in **Fig 25-6**, with its *EZNEC*

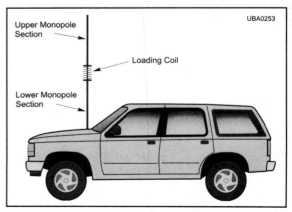

Fig 25-6 — HF whip antenna, with loading coil mounted on auto fender.

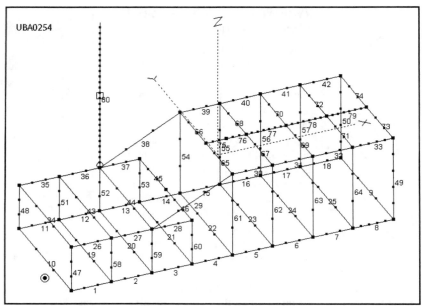

Fig 25-7 — *EZNEC* model of motor vehicle with antenna on front fender.

model, borrowed from a *QST* article. This model includes the car body itself acting as a grounding system, as shown in **Fig 25-7**.[1]

If a matching network is used, the requirements on the matching network are rather extreme. For example, an 8 foot whip antenna has an impedance at its base at 3.85 MHz of about $3.25 - j\,2395\ \Omega$. If you use a matching network with an inductor whose Q is 100, you would have losses in excess of 15 dB within the matching network itself. The loaded antenna, including such a matching network, will have similar losses.

It's interesting to look at the modeled results of various configuration of HF mobile antennas. **Fig 25-8** shows the elevation pattern for three cases. The first case (solid line) is a full-size monopole adjusted to be resonant at 3.85 MHz. While this is not too useful in a practical sense, it does indicate an upper efficiency limit. Next, insert an ideal, no-loss, loading inductor at the center of the 8 foot whip (dashed line), and finally use one with a realizable Q of 100 (dotted line). The maximum signal

intensity is summarized in **Table 25-1**.

Another result of the losses in the real inductor is shown in the SWR plots of the two loaded cases in **Fig 25-9** and **Fig 25-10**. Note that the ideal inductor results in a low resistive feed-point impedance at resonance and a very narrow bandwidth, narrower even than some modulation schemes. The low resistance requires additional matching elements, usually a shunt inductor at the base, in order to provide a matched feed to coax. By using a lossy real inductor, both problems disappear — along with about an additional 90% of the radiated power!

The 3.85 MHz example here is about the worst case for HF mobile antennas. At the other end of the HF range, 30 MHz, the 8 foot whip is a full-sized λ/4 monopole with performance comparable to a ground mounted antenna. Frequencies in-between perform in-between the two extreme cases, as you would expect, with antennas around half of full size (around 15 MHz for an 8 foot whip) not giving up much performance compared to full-sized antennas, as shown in **Table 25-2**.

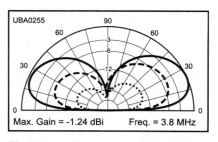

Fig 25-8 — Response of a full-sized λ/4 monopole (solid line) on vehicle, compared with an 8-foot whip using lossless center-loaded inductor (dashed line) and an 8-foot whip with a real inductor with a Q of 100 (dotted line).

Table 25-1

Transmitted Signal Intensity of Three 3.85-MHz Mobile Antenna Configurations.

Configuration	Signal Intensity (dBi)
Full size monopole	−1.24
Loaded 8 foot, ideal inductor	−5.85
Loaded 8 foot, real inductor	−15.36

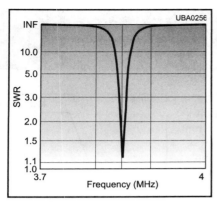

Fig 25-9 — 6.5 Ω SWR plot of 3.85 MHz center loaded antenna with lossless loading inductor.

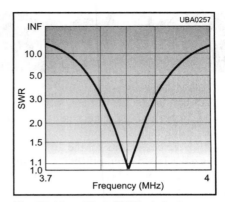

Fig 25-10 — 58 Ω SWR plot of 3.85 MHz center loaded antenna with real world loading inductor.

Table 25-2

Transmitted Signal Intensity of 8 Foot Center Loaded Mobile Antenna Compared to Full Size at Selected Frequencies.

Frequency	Full Size Height (feet)	Full Size Intensity (dBi)	8-Foot Signal Intensity (dBi)	Penalty (dB)
3.8 MHz	68.6	−1.24	−15.36	14.1
7.15 MHz	35.6	−1.21	−8.22	7.0
14.15 MHz	17.1	−0.2	−2.09	1.9
21.2 MHz	11.3	0	−.71	0.7
28.3 MHz	8.4	0.56	0.45	0.1

Water Borne Vehicles

Water borne vehicles often use technology for their antenna systems similar to that of land vehicles; however, they have a few significant advantages:
• Salt-water vessels have an almost ideal ground structure beneath them, and often a metal hull that is tightly coupled to it.
• Many watercraft are much larger than land vehicles in all dimensions. They also often include masts and other structures that can support taller HF antennas as well as providing a higher view for VHF/UHF systems.
• While on open water, they have much longer LOS distances than land vehicles do in all but the most flat of terrains.

Aircraft Antennas

Aircraft clearly are on top of the heap when it comes to LOS distances. On the other hand, they offer some of the most severe constraints for the antenna designer due to requirements that they not have protrusions that increase air resistance.

Traditional Aircraft Antennas

Early and slower aircraft used many systems in common with other vehicles — wire antennas for HF and whips for VHF/UHF. Many still employ such systems — a short mast forward on the fuselage secures one end of a wire that terminates on an insulator at the top of the tail's vertical stabilizer. This provides reasonable MF and HF coverage, if combined with a matching network inside the fuselage. Some slow-moving strategic communications platforms have even used vertically trailing wire antennas thousands of feet long to provide long range LF or MF communications.

Short VHF and UHF monopoles, constructed of thin wire have often been used on low to medium performance aircraft. If mounted below the fuselage, they provide coverage to a wide area.

Antennas for High-Performance Aircraft

High-performance aircraft provide a completely different perspective compared to their lumbering counterparts. In an aircraft world in which a rivet head is a no-no, any kind of protrusion is generally rejected. Fortunately, there are configurations that make use of slots, patches or other technology to allow a fully streamlined configuration. Such antennas are not generally added to the airframe, but are designed into the

skin of the fuselage as a part of the system-design process. I will discuss a few of the many solutions in current use.

Slot antennas are an ideal example of such a configuration. A slot in the skin of the fuselage can be driven by a waveguide from inside, or be backed by a box and coax fed, with the slot covered in insulating material with absolutely no indication from the outside that it is there — other than perhaps a "no step" admonition.

A variation of the slot is the *blade* antenna. In this case, rather than attempting to approximate an infinite ground sheet, specific tapered dimensions and offsets are used to provide coverage over a wide band of frequencies. This configuration is shown in **Fig 25-11**.

A vertical monopole with no height, at least from the perspective of drag, can be constructed using a flush disk as a capacitive loading element as shown in **Fig 25-12**. Again, with nonconductive material between the outside of the disk and the circular hole in the fuselage, the aircraft surface streamlining can be maintained. This antenna provides omnidirectional coverage with vertical polarization.

Some antennas, such as rotating radar dish antennas, are more difficult to streamline. While many aircraft that have extra space, place traditional dishes behind radomes that correspond to the shape of the aircraft, the more modern approach is to synthesize the beam shapes by using multiple slot or other streamlined antennas in phased arrays with the phase controlled by computer systems to determine the desired beam positioning.

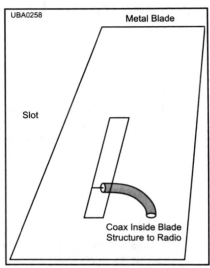

Fig 25-11 — Slot driven aircraft blade antenna.

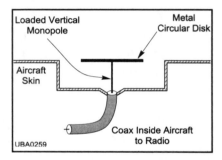

Fig 25-12 — Side view of flush disk antenna.

Chapter Summary

Antennas for use on vehicles are just adaptations of the same principles as antennas I have discussed previously. What makes vehicle antennas special is that they add additional constraints in terms of size — particularly height for land vehicles or air resistance for aircraft. The result is often a compromise between practicality and performance. Often the performance can be compensated for at the other end of the link, through higher antennas, additional power or sensitivity or special signal processing. Fortunately, there is usually a configuration that can be made to fit almost any application.

Notes

[1] S. Cerwin, "The Arch," *QST*, Jan 2008, pp 39-41.

Review Questions

25-1. What performance limitation makes MF vehicle receiving antennas easier to design than many others?

25-2. If heights as high as 12 feet can be accommodated by highway underpasses, what is the approximate lowest frequency that can use a full sized $\lambda/4$ monopole on a vehicle if the base of the monopole is mounted 2 feet above the roadway?

25-3. Why are slot, blade and disc antennas particularly suited for high-performance aircraft?

Chapter 26

Antenna Measurements

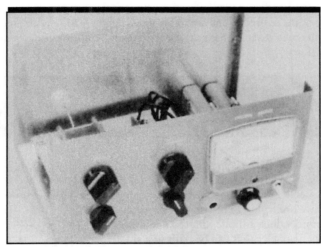

This homemade field strength meter can be used to map the response of transmitting antennas in different locations or directions.

Contents

Antenna Measurements

Measuring antenna characteristics and performance are important considerations in any antenna application or development project. Modeling is often much easier, quicker and better defined; however, there are always assumptions and limitations within the modeling process that can be best validated through actual physical measurements.

Measurement System Limitations

Unfortunately, it is not just modeling that has assumptions and limitations; measurements do as well. It is important to understand what some of these limitations are to be able to make valid judgments and draw appropriate conclusions from measured data. I will try to discuss these as I describe some of the various measurements that can be taken.

RF Voltage and Current

It is relatively straightforward to measure RF voltage, at least at relatively low frequencies. A series resistor, at least 100 times the value of the line Z_0 is connected to the line under load to minimize change to the conditions we are trying to measure. The resultant signal is rectified by a semiconductor diode, filtered and presented to a meter. A circuit of such a meter that can be built in a small metal box and placed in the line is shown in **Fig 26-1**.[1] It is also possible to build the components into a cigar tube with the resistor lead protruding and use it as an RF probe using a usual test meter as an indicator. Note that R1 and R2 must be *composition* resistors, not the more commonly used film type, which act more like inductors than resistors at radio frequencies.

The values of the components are shown in Fig 26-1 and are designed to make the meter read fairly linearly with RF voltage. You should establish reference points using a power or voltage meter of known accuracy for initial calibration of the voltmeter.

RF current can also be measured, by simply measuring the voltage across a small resistor in series with the line. To avoid changing the system SWR, the resistor should be both noninductive and it should have a value much less than that of the Z_0 of the line being measured. Note that in this case, both sides of the measurement system are at high RF potential with respect to the chassis of the metering circuit.

A type of RF ammeter called a *thermocouple* meter was in common use during WW II and samples are still frequently found. This meter, useful only through the HF region, uses a thermocouple, a device that converts temperature rise to voltage, to measure the RF current going through a resistance that is part of the thermocouple assembly. These types of RF ammeters are not commonly available as new products.

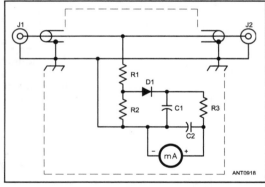

Fig 26-1 — Schematic diagram and parts list for an RF voltmeter for coaxial line. (Courtesy *The ARRL Antenna Book*.)
C1, C2 — 0.005 or 0.01 μF ceramic capacitor.
D1 — Germanium diode, type 1N34A or equivalent.
J1, J2 — Chassis mount coaxial sockets to match connectors in use.
M1 — Milliammeter, 0-1 mA or more sensitive.
R1 — 6.8 kΩ composition resistor. 1 W for each 100 W of RF through device.
R2 — 680 Ω, 1 W composition resistor.
R3 — 10 kΩ, ½ W composition or film resistor.

Standing Wave Ratio

Standing wave ratio (SWR) is one of the more straightforward measurements to make on an antenna system. SWR is a measure of how well the antenna, as a system load, matches the nominal system impedance. The usual way of measuring SWR is to separate the two waves that appear on a mismatched transmission line — the *incident* wave and the *reflected* wave. The incident wave represents the wave travelling from the signal source, which is usually a transmitter, towards the load, which is often an antenna. If the load impedance does not match the transmission-line characteristic impedance (Z_0), some fraction of the incident wave will be reflected back towards the transmitter. The power actually delivered to the load equals the difference so that $P_D = P_F - P_R$.

A common misconception is that the P_R term implies "lost power" in the system. Power can be lowered if the system cannot accommodate the actual load impedance seen at the transmitter. If the transmitter can perform properly, however, and if there are no other losses in the system, the reflected power is re-reflected at the source and heads back up the transmission line toward the load again. Of course, a real system cannot be totally lossless, and the losses do increase with SWR. Still if the transmitter can drive the resulting impedance seen at its output when there is an SWR, most of the power will reach the load.

Measuring SWR

The Direct Method

Incident and reflected waves will both exist along the length of the transmission line if there is SWR on that line. At any given point, a measurement of the voltage across the line will show the algebraic sum of the voltage from each wave. If there is no reflected wave (indicating a perfect match), the voltage for a lossless line will be constant along its length. On the other hand, if there is a reflected wave, it will add and subtract from the voltage of the incident wave as they propagate along the line. A direct measurement of voltage can be taken using a meter of the type shown in Fig 26-1. The SWR is defined as being equal to the ratio of the maximum voltage over the minimum voltage. Again, if there is no reflected wave, the voltage will be the same everywhere and the SWR will be 1:1.

To be sure that you have captured the extreme maximum and minimum values, you must take samples along a length equal to at least $\lambda/2$. This direct measurement requires that a non-loading voltage measurement, such as that just described, be made along the whole length. The object is to find the peak and minimum voltages, since their location will be a function of the phase angle of the load impedance. This method can be employed only if voltage sample points are available all along the line. This is not possible with regular coaxial cable, but is feasible with bare open wire line — if you can measure it without disturbing the values you are trying to measure.

At the high VHF to microwave range, it is possible to insert a section of *slotted line* into the system. A slotted line is just a precision rigid section of air-dielectric coaxial cable that has a slot through the shield, through which a voltage-measuring probe can be inserted. If the probe has a high impedance compared to the line Z_0, an accurate measure of the line voltage can be made at each point along the length of slot. If the slot is at least $\lambda/2$ long, the SWR can be determined directly from the ratio of maximum and minimum voltage readings.

While a slotted line (or even neon bulbs along HF open-wire line) can be used to measure SWR (or in the case of the neon bulbs to get a sense of the magnitude of SWR), the technique is no longer commonly used — except perhaps to demonstrate the concept in school laboratories. The most common method is to measure the forward and reflected power and compute SWR from those values.

SWR by Measuring Forward and Reflected Power

If a relatively short, in terms of wavelength, section of sampling transmission line is coupled to the main transmission line, inside the shield in the case of coax, a sample of the power going in either direction can be obtained. In **Fig 26-2**, a small sample (typically down 20 dB) of the signal in the main transmission line is obtained on the sampling line, which is terminated in a resistance at the far end. Reflected signals, returning from the output side of the main transmission line, are also coupled to the sampling line and go to the diode rectifier, which sends dc to a meter. With proper calibration, the reflected power can be read from the meter. Some systems have two sensors, one in each direction, so that forward and reflected power can be read simultaneously — others have a single sensor with metering circuit and the coupling loop is reversed to measure the power in the other direction. The

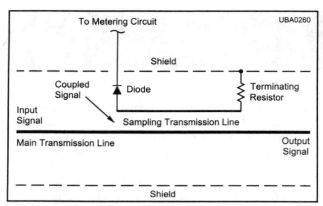

Fig 26-2 — Example of directional coupler used for measuring relative forward or reflected power.

coupling is a function of the fractional wavelength of the coupling line, so calibration is a function of frequency.

The forward and reflected powers do not show SWR directly, although some metering systems have scales calibrated to read directly in SWR. In most cases, the two power readings are noted and a calculation is made to determine SWR. This is easiest if done in two steps. The first step is to determine the magnitude of the reflection coefficient ρ. While the reflection coefficient is a complex quantity, the phase information is not needed here so you will just consider the magnitude of the quantity. This value is found as follows:

$$|\rho| = \sqrt{P_R / P_F}$$

(Eq 1)

The SWR is then found as follows:

$$SWR = \frac{1 + |\rho|}{1 - |\rho|} = \frac{1 + \sqrt{P_R / P_F}}{1 - \sqrt{P_R / P_F}}$$

(Eq 2)

This allows us to easily convert from forward and reflected power in watts or other units to SWR.

SWR Measurement Pitfalls and How to Avoid Them

SWR seems like a very straightforward item to measure; however, you are faced with a bit of a quandary. To perform the most accurate measurement, you need to measure at the load itself, which is usually an antenna. Placing measuring equipment, and even worse measuring people, at the antenna's feed point will prob-

ably result in a change in the impedance and thus the SWR. In addition, in many antenna installations, the antenna's feed point is at the top of a tall support structure — often not the best spot from which to be taking measurements.

The usual remedy for both problems is to measure the SWR at the bottom of the feed line. That's usually where the transmitter is and that's the part of the system that usually cares most about the SWR, so it seems like a best place to take the data. It can be, but it can also lead to errors in conclusions. If the line is lossless, the SWR at the bottom will equal that at the top. Unfortunately, real-world transmission line has losses, and the losses increase with SWR. Thus an SWR measurement at the bottom compares the full forward power with the attenuated reflection of the attenuated forward power that reaches the antenna.

Here's an example. Let's say you have a 100 W transmitter driving 100 feet of coax that has a loss of 3 dB (50% power loss) at some particular frequency. The antenna will see 50 W of power. But let's say 20% of the power is reflected due to the antenna being mismatched to the Z_0 of the transmission line. That will result in 10 W being reflected back toward the source. The 3 dB loss in the line results in 5 W showing up as reflected power at the transmitter at the bottom

of the line.

Table 26-1 summarizes the power at the two locations. Note the rather distressing result. A very acceptable measurement of an SWR of 1.6:1 at the bottom of the coax is the result of an unpleasant SWR of 4:1 at the antenna. In this example, our 100 W of power results in only 40 W radiated from the antenna — yet all of our measurements make us think we're doing well. Unfortunately, this example is not unusual, especially at the upper end of HF into the VHF range. If it happens at higher frequencies, it is usually more evident since nothing much ends up going in or out of the system! The sidebar discusses ways that this can be calculated, and perhaps avoided, through the use of software.[2]

Another problem that can confuse SWR measurements is if there are any stray common-mode currents present, such as those on the outside of the coax shield. This makes the transmission system a complex one, while the SWR instrumentation is happily measuring what is happening inside the coax, it only has a view into part of the total system. Thus, results are not accurate.

You can sometimes determine whether you have common-mode problems by observing the SWR as you move your hands along the outside of the coax. If everything is working as designed, there should be no change. However, if readings do change, stop taking the measurements — they will not be particularly meaningful. You will have to use appropriate measures, such as adding common-mode chokes or current baluns, to force currents to flow only on the inside of the coax, not on the outside of the coax's shield.

Table 26-1

Forward and Reflected Power as Seen at Each End of the Transmission Line

Measurement	Bottom of Cable	Top of Cable
Forward Power (W)	100	50
Reflected Power (W)	5	10
Indicated Reflection Coefficient	0.224	0.447
Indicated SWR	1.6	4.0

I Know What's Happening at the Shack — What's Happening at the Other End of my Feed Line?

If you want to find out — here's the easy way using TLW.

Joel R. Hallas, W1ZR
Technical Editor, QST

I'm told that one of the more frequent questions received by *QST's* "Doctor" has to do with folks wanting to determine the impact of transmission line losses on the effectiveness of their antenna system. These questions are often along the lines of "I measure an SWR of 2.5:1 at the transmitter end of 135 feet of RG-8X coaxial cable. My transceiver's auto-tuner can tune it to 1.1, but how can I tell what my losses are?" or "How much difference will I have if I have a tuner at the antenna instead of using the built-in tuner?"

These are important questions that almost every amateur operator is faced with from time to time. An approximate answer can be obtained by using the graphs found in any recent edition of The ARRL Antenna Book showing the loss characteristics of many transmission line types, plus adding in the effect of an SWR greater than 1:1. The SWR at the antenna end can be determined from the bottom end SWR and the cable loss. Using these graphs requires a bit of interpolation or Kentucky windage, but can result in useful data.

But There's an Even Better Way!

Packaged with each of the last few editions of *The ARRL Antenna Book* is a CD containing the pages of the whole Antenna Book as well as some very useful software. The program that I use almost daily is one written by Antenna Book Editor R. Dean Straw, N6BV, called *TLW* for Transmission Line for Windows.

TLW provides a very easy to operate mechanism to determine everything I usually need to know about what's happening on a transmission line. When you open the program, you are presented with a screen as shown in Figure 1. This has the values plugged in from the last time you used it, often saving a step. Let's take a quick tour of the inputs:

Cable Type — This allows you to select the cable you would like to analyze. A drop-down box provides for the selection of one of 32 of the most common types of coax and balanced lines. An additional entry is provided for User Defined Transmission Lines that can be specified by propagation velocity and attenuation.

Length — In feet or meters, your

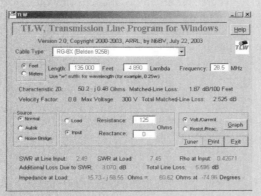

Figure 1 — The opening screen of *TLW*, **illustrating the process described in the article.**

choice.

Frequency — This is an important parameter when dealing with transmission line effects.

Source — This defines the form of the input impedance data. Generally, you can use NORMAL.

Impedance — The impedance can be specified as what you measure, resistive (real) and reactive (imaginary, minus means capacitive). This could come from your antenna analyzer at either end of the transmission line. Note, if you only know the SWR, not the actual impedance, all is not lost — see below.

Now for the Outputs

SWR — The SWR is provided at each end of the cable. This is an important difference that many people miss, important even with a moderate SWR at the transmitter end, as we'll see — the SWR at the antenna will be much higher due to the cable loss. With TLW, you instantly know the SWR at both ends, and the loss in the cable itself

Rho at Load — This is the reflection coefficient, the fraction of the power reflected back from the load.

Additional Loss Due to SWR — This is one of the answers we were after.

Total Loss — And this is the other, the total loss in the line, including that caused by the mismatch.

But Doctor, What if I can Only Measure the SWR —

Not the Actual Impedance?

Often the only measurement data available is the SWR at the transmitter end of the cable. Because the losses are a function of the SWR, not the particular impedance, you can just put in an

arbitrary impedance with that same SWR and click the INPUT button. An easy arbitrary impedance to use is just the SWR times the Z_0 of the cable, usually 50 Ω. For example, you could use a resistance of 125 Ω to represent an SWR of 2.5:1. This is what we've done in Figure 1, using 135 feet of popular Belden RG-8X.

The results are interesting. Note that the 2.5:1 SWR as seen at the radio on 28.5 MHz results from a 7.45:1 SWR at the antenna — perhaps this is an eye-opener! Note that of the 5.6 dB loss, more than half, or 3.1 dB, is due to the mismatch. Note that if we used something other the actual measured impedance, we can't make use of the impedance data that TLW provides. We can use the SWR and loss data, however, but that's probably what we wanted to find out.

We can now do some "what ifs." We can see how much loss we have on other bands by just changing the frequency. For example, on 80 meters, with the same 2.5:1 at the transmitter end, the SWR at the antenna is about 3:1 and the loss is slightly more than 1 dB. We could also plug in an impedance calculated at the antenna end and see what difference other cable types would make. For example, with the same 28.5 MHz SWR of 7.45 at the antenna and 135 feet of 1/2 inch Andrew Heliax, we will have a total loss of 1.5 dB at 28.5 MHz. Note that the SWR seen at the bottom will now be 5.5:1 and our radio's auto-tuner might not be able to match the new load.

But Wait There's More!

You can also click the GRAPH button and get a plot of either voltage and current or resistance and reactance along the cable. Note that these will only be useful if we have started with actual impedance, rather than SWR.

Pushing the TUNER button results in a page asking you to select some specifications for your tuner parts. TLW effectively designs a tuner of the type you asked for at the shack end of the cable. It also calculates the power lost in the tuner and gives a summary of the transmitted and lost power in watts, so you don't need to calculate it!

When you've finished, be sure to hit the EXIT button, don't just close the window. Otherwise TLW may not start properly the next time you want to use it.

Field Strength

Field strength is a quantity that is easy to measure, but that can be difficult to interpret. A basic field-strength meter (FSM) can be as simple as an untuned crystal radio receiver with a short antenna connected to a meter movement, as shown in **Fig 26-3**.

For most antennas, the long-range performance is what you'd like to measure. For a multielement array that exhibits a good deal of gain, such as a long-boom Yagi, the field-strength meter must be physically far away from the array, well into the *far field*. The far field boundary is a function of the size and complexity of the array, as well as the wavelength. The distance to the boundary, R, can be approximated by the following:

$$R = \frac{2L^2}{\lambda} \qquad \text{(Eq 3)}$$

where L is the largest antenna dimension, and λ is the operating wavelength, all in the same units. This expression is valid for L > λ. Note that for the case of a 1λ boom Yagi, L = λ, and Eq 3 reduces to:

R = 2 λ.

Making Field-Strength Measurements

A meter such as the one in Fig 26-3 can be used to provide a

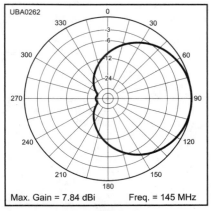

Fig 26-4 — Elevation pattern of three element Yagi in Chapter 19, in free space.

relative indication of field strength. There are difficulties associated with attempting absolute measurements. For example, a field-strength meter can be a good tuning aid to determine antenna, transmission line or transmitter settings that provide maximum signal strength to a particular location, or to adjust phasing lines for a null in a specific direction. Since the scale is not necessarily linear, comparison between signals is best done with the meter at the same point and a variable attenuator or calibrated transmitter power setting used to achieve the same response.

In spite of a number of limitations and caveats with such devices, as well as the general field-strength issues to be discussed, there are useful measurements that can be obtained with such a simple device. For example, to measure an antenna's front-to-back ratio at the location of the FSM, first point the back of the antenna towards the FSM and adjust the power output for a solid reading. Turn off the transmitter and make sure the meter reads zero. Other transmitters in the area can make the use of an untuned FSM unusable!

Turn the transmitter back on and make note of the transmitter power. Turn the antenna so the front of the antenna is facing the FSM, with everything else the same. Reduce the transmitter power until the same reading is obtained and measure the power. The ratio of the recorded powers is the F/B.

Another approach to obtaining relative measurements is to use the antenna under test as a receiving antenna. In this case you use a nearby steady signal source, along with a stable receiver. Many receivers incorporate a signal-strength indicator. While some have calibrations indicating signal strength, few are accurate, as I've indicated before. If a calibrated signal generator is at hand, the meter indications, or possibly the automatic gain control (AGC) voltage

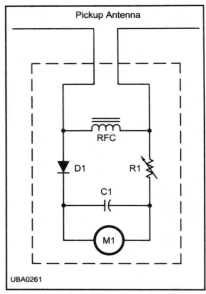

Fig 26-3 — Schematic diagram and parts list for a simple RF field strength meter.
C1 — 0.005 or 0.01 µF ceramic capacitor.
D1 — Germanium diode, type 1N34A or equivalent.
M1 — Milliammeter, 0-1 mA or more sensitive.
R1 — Rheostat for setting meter sensitivity. Value should be about 100 times the meter resistance.
RFC — RF choke with at least 10 kΩ impedance at operating frequency. Alternately, a 10 kΩ or higher value composition resistor can be used.

can be used to directly compare signals for F/B or to make direct A-to-B comparison measurements.

If a calibrated step attenuator is available, an even better way is to take a reference reading on the weaker signal with the attenuator at 0 dB, and then click in attenuation until the stronger signal is reduced to the same reading. The value of the attenuator then is the same as the difference between readings.

Field-Strength Measurement Pitfalls and How to Avoid Them

The biggest challenge facing those

trying to evaluate antenna performance is generally separating the antenna from its environment. In this case, what this means is anything around the antenna that will modify its performance. The ground is often the biggest hurdle, however, nearby antennas, aerial cables and even metal rain gutters can all modify the way an antenna performs.

As an example, look at the 145 MHz Yagi examined in Chapter 19. There I examined the antenna in free space to avoid just the kind of problem being addressed here. The free-space elevation pattern is shown in **Fig 26-4**. This is the pattern associated with the antenna devoid of any external influences. Compare that to the pattern at a typical height of 40 feet above real ground shown in **Fig 26-5**. Note that the maximum

gain has increased by a bit less than 6 dB due to ground reflections; however, the gain varies by as much as 20 dB as you make slight changes in elevation. Compounding this height effect is that the FSM antenna will also exhibit the same performance, just as any antenna will if it is mounted more than about λ/2 over ground.

In an antenna range, the usual practice is to adjust both the antenna under test and the sense antenna up and down to make sure each is at the peak of response. A reference antenna, such as a dipole, with a known gain, then replaces the antenna under test. The process is repeated for the dipole and then the gain difference between the two can be stated.

For this kind of measurement, it is also very useful to model the antenna beforehand to gain insight into how

quickly performance changes with azimuth angle. This will demonstrate the resolution required in adjustments.

Some parameters are more sensitive to ground reflections than others. For example, **Fig 26-6** is the azimuth plot of the antenna over ground at the elevation angle of its first peak at 2°. **Fig 26-7** is the azimuth plot with the elevation set near the first null at 5°. Note that while the forward gain difference is more than 20 dB, both plots have a F/B of 31 dB, within a few tenths of a dB, suggesting that front-to-back measurements can be made with little concern about where the antenna is in the elevation pattern, a fact I implicitly took advantage of earlier.

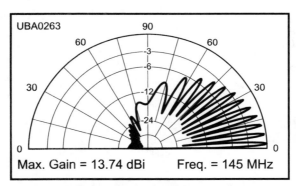

Fig 26-5 — Elevation pattern of three element Yagi 40 feet above typical ground.

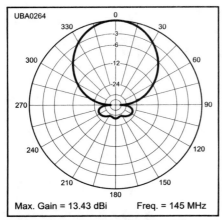

Fig 26-6 — Azimuth pattern of three element Yagi 40 feet above typical ground at 2° elevation.

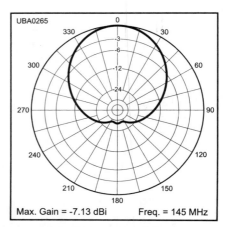

Fig 26-7 — Azimuth pattern of three element Yagi 40 feet above typical ground at 5° elevation.

Chapter Summary

Many systems are available to measure antenna operation and performance; however, their effective use requires an understanding of how to separate the antenna from its environment. This is particularly critical for gain measurements, but can impact all types of measurements. It is also critical to understand what it is that is being measured and how measurements in one location can be properly or improperly used to determine what is happening.

Review Questions

26-1. How can current and voltage measurements be used to determine power on a transmission line? What precaution must be taken, particularly if the SWR is greater than 1:1 to make the calculation valid?

26-2. Under what conditions can't the SWR at the bottom of a transmission line be directly used to determine antenna SWR?

26-3. What are the limitations that affect the ability to use field-strength measurements for absolute gain determination?

Notes

[1]R. D. Straw, Editor, *The ARRL Antenna Book,* 21st Edition. Available from your ARRL dealer or the ARRL Bookstore, ARRL order no. 9876. Telephone 860-594-0355, or toll-free in the US 888-277-5289; **www.arrl.org/ shop/; pubsales@arrl.org.**

[2]J. Hallas, W1ZR, "I Know What's Happening at the Shack -- What's Happening at the Other End of my Feed Line?" *QST,* Feb 2007, p 63.

Appendix A

Getting Started in Antenna Modeling with EZNEC

Contents

Introduction

You may well wonder how someone can determine just how well and in which directions an antenna will radiate. This is an old problem traditionally settled using an *antenna range*, which consisted of a large open space with big towers and expensive measuring equipment so that the actual directivity and gain of an antenna could be determined.

Of course, there have always been theoretical methods to determine antenna performance, but these often couldn't take into account some important factors. Key elements often missing in theoretical analyses have been the effects of ground reflections and antenna interactions with other nearby objects, such as support structures, or vehicle structure for mobile, airborne or shipboard antennas. Fortunately, in our twenty-first century computer-oriented society, we have software tools that help us avoid the need to climb towers.

The radiation from an antenna can be modeled by dividing up the antenna into a number of segments, and then adding up the radiation from each of these individual segments as seen at a distant point. The computer must take into account both magnitude and phase from each segment, and can determine the resultant as the energy combines at some distant location, at any desired angle with respect to the antenna.

Antenna-simulation programs make the assumption that individual segments are small enough so that the current leaving a segment is the same as the current entering it. Other structures, such as other antennas, guy wires or towers, not connected to the antenna itself can also be modeled, and the currents induced in them will be part of the overall calculation.

The resulting model can be an effective tool to predict how a particular antenna will function, without having to actually construct and then measure the way it works. As with any simulation, your mileage may vary, but under most circumstances modeling provides at least a good starting point.

EZNEC

One popular antenna-modeling program is named *EZNEC* (pronounced *easy-nec*). The "NEC" part of the name comes from the core calculation engine, the *Numerical Electromagnetics Code*, a powerful antenna-analysis tool that forms the basis of a number of antenna analysis programs. The "EZ" part of the name comes from the fact that this *Windows* implementation is indeed easy to use. *EZNEC* is available from its developer, Roy Lewallen, W7EL at **www.eznec.com**. The program is available in a number of versions, including a free demo version with no time limit. The demo is restricted in size to 20 segments, enough to get a feel for simple antennas such as a dipole.

Getting Your Feet Wet With EZNEC

If you follow along in this Appendix, you will get an understanding of how to use *EZNEC* to model antennas and obtain results very quickly. It is important to note that as antennas get more complicated, or get close to the ground, a number of complications make accurate results more difficult to obtain. Look at the information on the *EZNEC* Web site, or at other sources such as **www.cebik.com**, to get a sense of the limits in modeling.

To use *EZNEC*, you might start out with a sample antenna definition (some are supplied with the program). Or you might decide to enter the physical dimensions of your antenna in X, Y and Z coordinates. You also must specify wire gauge or diameter and pick a segment quantity. As mentioned above, the free demo version has a limit of 20 segments, while the basic version has a limit of 500 segments. These values are all entered on the WIRES tab.

On the SOURCES tab you specify which wires will be connected to source(s), and the location where a source connection will go in terms of percentage distance from one end of a wire. Pick the type of ground you want from the GROUND tab choices. That's all it takes to model an antenna.

To calculate some results, you might select the SWR tab and give the program a range of frequencies over which to determine the antenna's impedance and SWR (the default is 50 Ω, but you may specify other impedances). Most folks will want to see a plot of their antenna's pattern. You may select azimuth or elevation plots on the PLOT TYPE tab, and then you would click on FF PLOT to see the results. It takes less time to do it than talk about it.

Of course, there are some refinements and details that make such modeling even more useful, but I expect you will find out about them as you need them. If you want to determine the effect of changing the length of a wire, you can adjust the dimensions in the WIRES tab and run the calculations again. If you want to find how your antenna works at another frequency, just type in another frequency in the FREQUENCY tab and click FF PLOT again.

Using EZNEC to Model Some Sample Antennas

I have put together some *EZNEC* models of a simple antenna to serve two purposes — first, to show how easy it is to use, and second to illustrate some antenna principles that we have discussed in an abstract way earlier.

A Dipole in Free Space

The first antenna I will model is a horizontal half-wave dipole in free space. I have picked a frequency of 10 MHz (a 30 meter wavelength). I will use the antenna configuration shown in **Fig A-1**. The WIRES entry table that I created is shown in **Fig A-2**.

Note that I have picked a particular length so that the antenna will be resonant at almost exactly 10.0 MHz. Exact resonance is not really necessary, but this does represent a half-wave resonant dipole. I selected a height (the Z coordinate) to be half a wavelength above ground. Note that for "free space" the ground height doesn't matter, but you must specify some height.

I selected AWG #14 wire copper wire for the material. I could have specified it as a wire size (with the # symbol) or put in a number indicating diameter in inches. It is possible to set the table up in metric or English units, by the way. I selected a number of segments (19) to fit within the free demo. The program will give a warning message if you select a number of segments that result in too small or large a segment length. Keeping in mind the 500 segment limit (20 for the free demo), one way to gauge the size is to temporarily double the number and see if the results change very much. If not, your original selection is likely to provide valid

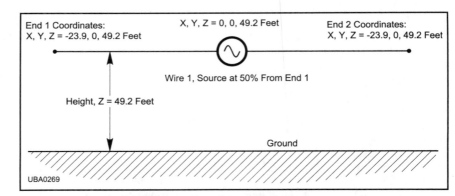

Fig A-1 — **Physical configuration of modeled antenna.**

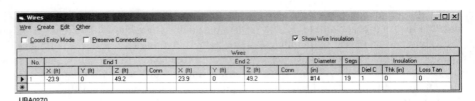

Fig A-2 — *EZNEC* WIRES **data entry screen.**

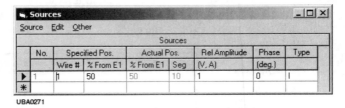

Fig A-3 — *EZNEC* SOURCES **data entry screen.**

results.

Now you've got a model and the *EZNEC* main window should look like **Fig A-4**. By clicking the SWR button, you would bring up the screen in **Fig A-5** on which you can specify the range and resolution you would like for an SWR plot. **Fig A-6** shows the resulting plot of SWR around the resonant frequency. Note that by putting the cursor on the 10 MHz axis, the data shown is for 10 MHz. I could select any other frequency in the range. The readout starts out with the data at the low frequency limit.

Note also that I could use this

window to "trim" the antenna to be resonant, with a reactance close to zero Ω. I would make the dimensions longer if the reactance were negative, and shorter if the reactance is positive. I can continue in this fashion until I got quite close to zero. Note also that my final length, 29.9 feet is close to resonance and that I could have gotten even closer by using more decimal places. Going beyond about 1/10 of a foot for an HF wire antenna is a level of precision that works in the model, but is not realistic for my real-world ruler and wire cutters. The message is that the model is fine as

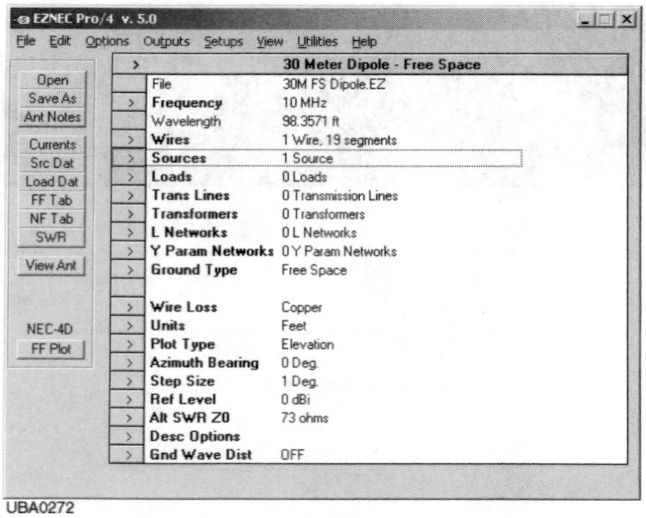
EZNEC Pro/4 v. 5.0
File Edit Options Outputs Setups View Utilities Help
Open
Save As
Ant Notes
Currents
Src Dat
Load Dat
FF Tab
NF Tab
SWR
View Ant
NEC-4D
FF Plot
30 Meter Dipole - Free Space
File 30M FS Dipole.EZ
Frequency 10 MHz
Wavelength 98.3571 ft
Wires 1 Wire, 19 segments
Sources 1 Source
Loads 0 Loads
Trans Lines 0 Transmission Lines
Transformers 0 Transformers
L Networks 0 L Networks
Y Param Networks 0 Y Param Networks
Ground Type Free Space
Wire Loss Copper
Units Feet
Plot Type Elevation
Azimuth Bearing 0 Deg.
Step Size 1 Deg.
Ref Level 0 dBi
Alt SWR Z0 73 ohms
Desc Options
Gnd Wave Dist OFF
UBA0272

Fig A-4 — *EZNEC* main screen showing a summary of our model parameters.

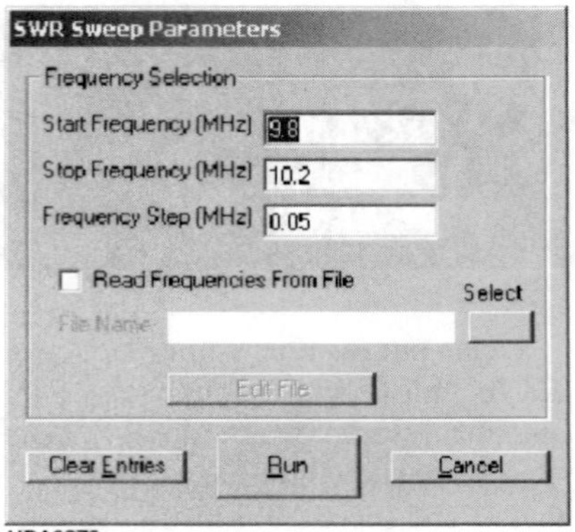
SWR Sweep Parameters
Frequency Selection
Start Frequency (MHz) 9.8
Stop Frequency (MHz) 10.2
Frequency Step (MHz) 0.05
Read Frequencies From File
Select
File Name
Edit File
Clear Entries
Run
Cancel
UBA0273

Fig A-5 — *EZNEC* SWR parameter entry screen.

far as it goes, but don't let it result in unattainable values!

If I were actually making an antenna like this (and I have many times), I would use the model as a guide and start with perhaps an extra 6 inches of antenna length on each side. I'd raise the antenna to its operating height temporarily, measure the SWR over the frequency range I wanted, lower the antenna and shorten the ends a little at a time until I achieved the results I wanted.

If you want to adjust your model to indicate the SWR at a different Z_0, you merely change the Alternate Z_0 on the main *EZNEC* window. For example, you might change to 73 Ω, the value that the antenna shows at resonance. The results would be shown in **Fig A-7**.

You could also use the model to predict the SWR over a wider frequency range. You just need to make another pass and change the SWR entry frequency limits. An example is shown in **Fig A-8**.

Note that this highlights the fact that most antenna types have multiple resonances, as seen at slightly above the third harmonic of the 10 MHz half-wave resonance at 30.5 MHz. Some antenna designs take advantage of this effect.

The antenna patterns of your dipole are shown below for azimuth (**Fig A-9**) and elevation (**Fig A-10**). Note that the length of the line from the center of the plot to the pattern at a particular angle shows the signal strength at that angle. In this case of a dipole in free space, the elevation pattern is uniform all the way around the antenna, meaning that the elevation pattern is "omnidirectional."

The azimuth pattern shown is typical of what we would expect of a dipole — maximum radiation broadside to the antenna and no radiation from the ends. As expected, the maximum broadside radiation is about 2 dBi, 2 dB above an isotropic antenna.

Down to Earth — the Dipole Model over Real Ground

The last section dealt with an ideal dipole in free space. While space is still pretty free, it's hard to get to and most dipoles are fabricated closer to the earth! A feature of *EZNEC* is the ability to easily move between free space, perfect ground (as in an infinite gold sheet) and real ground. By just clicking on the GROUND tab of the main *EZNEC* window you are offered a choice. I have done that with this model, as shown **Fig A-11**.

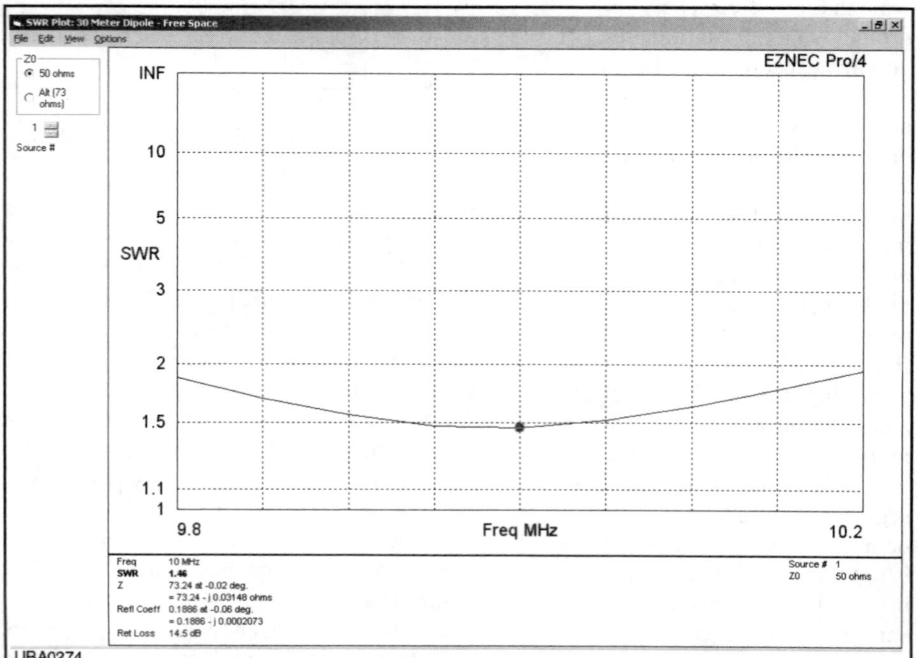
SWR Plot: 30 Meter Dipole - Free Space
File Edit View Options
Z0
50 ohms
Alt (73 ohms)
Source #
EZNEC Pro/4
INF
10
5
SWR
3
2
1.5
1.1
1
9.8
Freq MHz
10.2
Freq 10 MHz
SWR 1.46
Z 73.24 at -0.02 deg.
= 73.24 - j 0.03148 ohms
Refl Coeff 0.1886 at -0.06 deg.
= 0.1886 - j 0.0002073
Ret Loss 14.5 dB
Source # 1
Z0 50 ohms
UBA0274

Fig A-6 — *EZNEC* SWR plot for modeled antenna in free space with 50-Ω reference.

You can accept the *EZNEC* default ground parameters or insert actual ones depending on your knowledge of local soil conditions.

Fig A-12 shows the resulting SWR plot. While it has a similar shape to that of the free-space plot, you can see that the resonant frequency has shifted. This is a typical, in that changing almost anything about an antenna causes change in the performance. This is one reason why many antennas have adjustable element lengths to compensate for changes in local conditions.

The elevation plot in **Fig A-13** is very different from Fig A-10, the plot for the free-space antenna. Fig A-13 is for a dipole placed a half wave above flat ground. It dramatically shows the result of a reflected signal from the flat ground below. For a horizontally polarized antenna, the reflection is out of phase with the incident wave. At a height of a half wavelength above ground, the radiation cancels in the upward direction at very low takeoff angles, for example, at the horizon. At very high elevation angles, radiation cancels, for example, straight up.

As the plot data indicates, the maximum signal appears at an elevation angle of 28° with the dipole mount a half wavelength high.

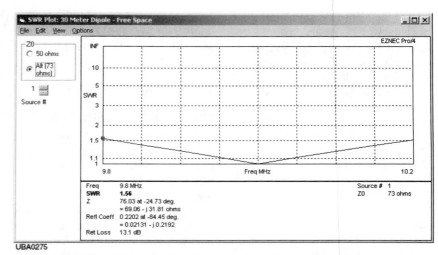

UBA0275

Fig A-7 — Data of Fig A-6 with reference impedance (Z$_0$) changed to 73 Ω.

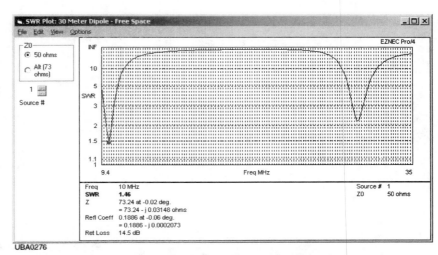

UBA0276

Fig A-8 — Data of Fig A-6 with expanded frequency range.

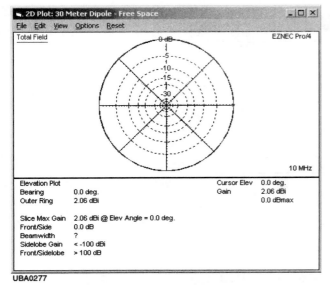

UBA0277

Fig A-9 — *EZNEC* graphical output (FFPLOT) with PLOT TYPE set to ELEVATION.

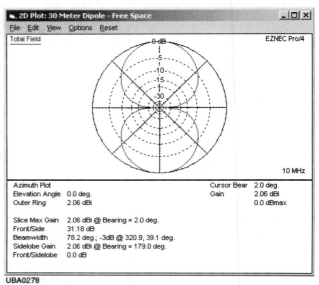

UBA0278

Fig A-10 — *EZNEC* graphical output (FFPLOT) with PLOT TYPE set to AZIMUTH, elevation angle set to 0° (horizon).

In general, as the antenna is raised higher above the ground, the major radiation lobe is lowered, but it never quite reaches the horizon, no matter how high it is placed.

The plot in **Fig A-14** provides the azimuth pattern at a 28° elevation angle. Note that *EZNEC* allows you to specify any elevation angle for the azimuth plot. I selected the elevation that yielded the maximum signal, but you may be interested in the signal at other takeoff angles. In Fig A-14 *EZNEC* has selected the azimuth with the strongest signal for the detailed readouts at the bottom of the plot. As with all *EZNEC* plots, by merely clicking on another point on the plot, you can select the readout for that point. In **Fig A-15**, I have selected the data at 45° off axis, just by clicking on that point of the curve. This is a very powerful tool!

Going Further with *EZNEC*

The various forms of dipoles we've looked at here just scratch the surface of *EZNEC*'s capabilities. Additional elements can be added just by defining additional "wires" separate from the first. For parasitic elements, that's all there is to it. For additional driven elements, multiple sources can be defined or the elements can be connected using transmission lines, defined in their own parameter screen.

Most outputs are available in tabular form so they could be input into other programs for further analysis. As with all such analysis tools, more complicated antenna models can have pitfalls, and it is best to read the *EZNEC* documentation to understand how far limitations can be pushed before the results suffer.

In any event, try different models of the antennas shown in the book or that you think might provide interesting results. It is much easier, faster and cheaper than building them in the back yard!

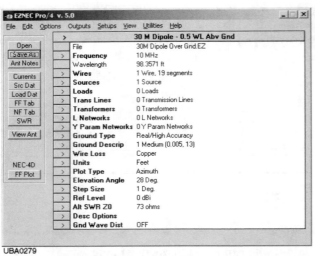

UBA0279

Fig A-11 — *EZNEC* main screen after selecting REAL/ HIGH ACCURACY ground in place of FREE SPACE.

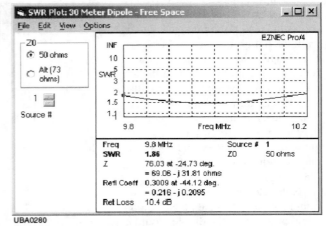

UBA0280

Fig A-12 — *EZNEC* SWR plot for modeled antenna over real ground with 50-Ω reference.

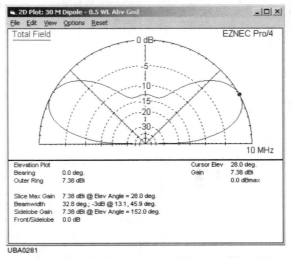

UBA0281

Fig A-13 — *EZNEC* graphical output (FFPLOT) with PLOT TYPE set to ELEVATION for model over real ground.

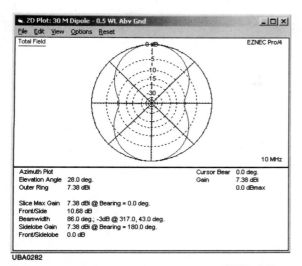

Fig A-14 — *EZNEC* graphical output (FFPLOT) with PLOT TYPE set to AZIMUTH, elevation angle set to 28° for model over real ground.

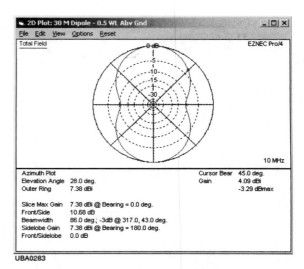

Fig A-15 — Same plot as Fig A-14, except cursor is set to a 45° azimuth angle to determine value is at that bearing.

Using Decibels in Antenna Calculations

Contents

Introduction

Much of the technical data found in the usual antenna references from ARRL and others is expressed in decibels, generally written as *dB*. Once you become familiar with the idea you will likely find that decibels make comparisons easier. Decibels also avoid having to use the very large or small numbers that often appear in system calculations.

What Are Decibels All About?

Decibels are just a way of expressing ratios, usually power ratios. If you are looking at the gain of an amplifier stage, the pattern of an antenna or the loss of a transmission line you are generally interested in the ratio of the power out to the power in. In antenna work, you are often concerned with the ratio of the power in front of a beam antenna to that coming from the back. These are some of the places where you will find the results expressed in dB.

Decibels are a logarithmic function. An important feature of logarithms is that you can perform multiplication by adding logarithmic quantities instead of multiplying them. Similarly, you can divide numbers by subtracting the same manner. This becomes a benefit if you are dealing with multiple stages of amplification and attenuation — as you often are doing in radio systems. In a radio receiver you are snatching a miniscule signal from the air, and then amplifying and processing it so you can hear it from a loudspeaker. Instead of having to multiply and divide at each stage to keep track of the progress of our signal processing — often with signal levels with many zeros to the right of the decimal point — you can just add up all the dB and determine the total gain in the system.

How Do We Compute Decibels?

The *deci* in decibels refers to a factor of $\frac{1}{10}$, as in *deciliters* for 1/10 of a liter, while the *bel* relates to the idea of a logarithmic ratio, originally used to define sound power. The bel was named after Alexander Graham Bell, the inventor of the telephone.

• To convert a power ratio into decibels, just:
1. Find the base 10 logarithm of the power ratio.
2. Multiply by 10.

The equation for this is:

Power ratio in decibels = $10 \, Log_{10} \, (P_2/P_1)$

For example, if we have an amplifier with a power gain of 275, you find the logarithm of 275 (see below, if you don't do logs in your head), which is 2.439. Multiply by 10 and the result is that a power gain of 275 can be represented as 24.39 dB.

• To convert decibels to a power ratio, we do the opposite:
1. Divide by 10.
2. Find the base 10 antilog of the result.

Note that the base 10 antilog of a number is just 10 raised to the power of the number. This is also something you probably don't do in your head; so let's see how you can easily perform the computations.

Understanding a few characteristics of logs will help avoid problems interpreting results. Note that a gain of 0 dB, means that there is no change to the signal — not that the signal has vanished! The other important fact is that a power ratio of less than one (a loss rather than a gain) results in a negative number in decibels.

Enter the Windows Scientific Calculator

In the *very* old days, engineers and technicians used printed tables to make accurate logarithmic calculations. They used mechanical *slide rules* if three significant digits was sufficient precision.[1] Starting around 1970, scientific calculators became available. Initially they were expensive typewriter sized devices that were typically shared within an engineering department. Within a few years pocket-sized units became available for less than $200, and then everyone could make calculations to a precision of nine significant digits, whether warranted by the data precision or not. Tables and slide rules were relegated to the pages of history, along with spark transmitters.

The arrival of the reasonably priced personal computer seemed to push the fancy scientific calculator out of sight only about 10 years later. Unfortunately, for many functions a calculator may be a better choice and decibel calculations may be one of those. If you have a suitable scientific calculator, it should easily do your calculations. Not all have an **ANTILOG** button, but if not, they will likely have a button that says **X^Y**, which can be used as above. If you don't have a handheld calculator, you may not know that there is a very capable one included as an "accessory" within the Microsoft *Windows* operating system! Just click START, then all programs, then accessories.

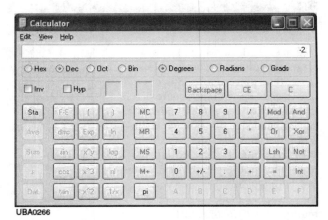

UBA0266

Fig B-1 — Screen shot of the *Windows* scientific calculator ready to find the power loss of –2 dB.

You should find an icon for the *Windows Calculator*. You could open it, but why not drag it onto your desktop first?

On first opening, you may find a four-function grocery store type calculator. Have no fear; just click on **VIEW,** then **SCIENTIFIC** to get the one you want. It should look like **Fig B-1**.

Give it a Test Drive

Let's say you have a mismatched coax cable with a loss of 2 dB. You may want to know how much of the 100 W generated by your transmitter actually reaches your antenna. Remember, a 2 dB loss is a "gain" of –2 dB! Using your *Windows Calculator*, either hit **2** on your keyboard, or:
- Click on the **2** on the calculator keypad. Then:
- Click on the ± key, the display should show **–2**, as in Fig B-1.
- Click on the **/** key to select the divide-by operation, then

click enter. The display should show **–0.2**.
- Enter the digits **1** and **0** for the number 10.
- Click on **X^Y** to raise 10 to a power.
- Click on the decimal point, then the digit **2** and ± and click enter.

The display should show 0.6309573444 (a number with many digits), which is about 0.63. That is the fraction of your power left after a 2 dB loss. That means of the 100 W you transmit, your antenna sees 63 W and 37 W is heating your transmission line.

Using Decibels to Represent Voltage and Current

The same decibels can be used to represent voltage or current ratios, rather than power ratios. Since power goes up with the square of the voltage or current, assuming the same impedance, the voltage or current ratios must be squared as well. With logarithms, ratios are squared merely by multiplying by two. Thus everything works the same way as for power calculations, except we multiply by 20 instead of 10:

Voltage ratio in decibels = 20 Log_{10} (V_2/V_1)

Thus an amplifier with a power gain of 20 dB has a power gain of 100 and a voltage gain of 10. Again, these values are easy to compute as described above. It is also easy to set up a spreadsheet to make the calculations, especially if you do them frequently. **Fig B-2** shows an *Excel* spreadsheet I set up for this purpose. The formulas are shown in **Fig B-3**, should you wish to duplicate it and check my work.

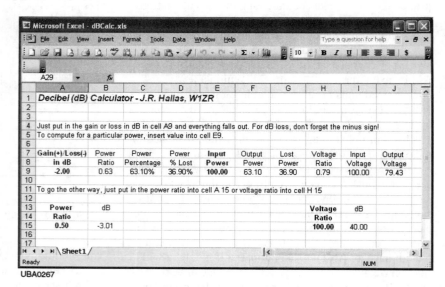

UBA0267

Fig B-2 — Screen shot of the author's decibel-calculating spreadsheet.

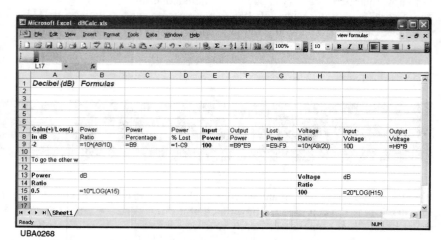

UBA0268

Fig B-3 — Screen shot showing equations behind the calculations of Fig B-2.

Decibel Calculations with Special References

While the general use of decibels is as a ratio of input to output, they are also used to represent particular power or voltage levels by defining the base power or voltage at a particular level. In this case, the dB units will have a suffix indicating the reference. For example, if we want to indicate power as compared to 1 W, we would use the symbol dBW to mean just that. For example, a signal level of 13 dBW would indicate the power of 200 W. Other common references are shown below:

- dBd — decibels of gain with respect to a dipole antenna in its preferred direction. dBi — decibels of gain with respect to that of an ideal isotropic antenna.
- dBmV — voltage level in decibels compared to a millivolt.
- dBμV — voltage level in decibels compared to a microvolt.
- dBm (dBmW) — power level in decibels compared to a milliwatt.
- dBμW — power level in decibels compared to a microwatt.

For example, if we had an amplifier with a gain of 30 dB and applied an input signal of 10 dBm, we would have an output power of 40 dBm or 10 dBW, which is 10 W. Note that the amplifier gain is in dB, a unitless ratio, while the input and output levels are in dBm or dBW, representing particular powers.

Notes

[1]See, for example, *Standard Mathematical Tables*, CRC Press, any edition. In addition to tables of logarithms and trigonometric functions, this book includes many handy formulae from geometry, trigonometry and calculus.

Notes

Notes

Notes

Index

FEEDBACK

Please use this form to give us your comments on this book and what you'd like to see in future editions, or e-mail us at **pubsfdbk@arrl.org** (publications feedback). If you use e-mail, please include your name, call, e-mail address and the book title, edition and printing in the body of your message. Also indicate whether or not you are an ARRL member.

Where did you purchase this book?
  ☐ From ARRL directly   ☐ From an ARRL dealer

Is there a dealer who carries ARRL publications within:
  ☐ 5 miles   ☐ 15 miles   ☐ 30 miles   of your location?   ☐ Not sure.

License class:
  ☐ Novice   ☐ Technician   ☐ Technician with code   ☐ General   ☐ Advanced   ☐ Amateur Extra

Name _____

ARRL member?   ☐ Yes ☐ No

Call Sign _____

Daytime Phone   (   ) _____   Age _____

Address _____

City, State/Province, ZIP/Postal Code _____

If licensed, how long? _____   e-mail address: _____

Other hobbies _____

_____

Occupation _____

From _____

_____

_____

EDITOR, BASIC ANTENNAS
ARRL—THE NATIONAL ASSOCIATION FOR AMATEUR RADIO
225 MAIN STREET
NEWINGTON CT 06111-1494

— — — — — — — — — — — — please fold and tape — — — — — — — — — — — — —